LABORATORY EXPERIMENTS FOR

Introduction to Chemistry

LABORATORY EXPERIMENTS FOR

Introduction to Chemistry

Fifth Edition

T.R. Dickson
Cabrillo College

John Wiley & Sons
New York Chichester Brisbane Toronto Singapore

ISBN 0-471-85048-9
Printed in the United States of America

10 9 8 7 6 5 4 3

Preface

Chemistry is an experimental science and laboratory experience is a necessary part of the study of chemistry. Many important concepts are vitalized in laboratory exercises. Furthermore, much nonverbal learning is accomplished by the student through laboratory work. This laboratory manual is designed for a first course in chemistry. It is intended for students who have had no laboratory experience. Along with vitalizing important chemical concepts the manual emphasizes laboratory techniques, procedures, and safety.

This manual provides flexibility in the laboratory experience. A variety of experiments is included so that the instructor can select the preferred laboratory sequence. Some experiments stress the observations of the properties and reactions of chemicals while others emphasize the collection of data used to carry out calculations. The students are provided with detailed directions but are also challenged by situations in which they must draw conclusions, describe their observations, identify unknowns, and carry out calculations.

The fifth edition of this manual has been revised using the experience of students and the advice of instructors who have used it. An effort has been made to eliminate or minimize the use of potentially hazardous and toxic chemicals. Acids, bases and hydrocarbons are the only hazardous chemicals used. (Acetic anhydride is required in the last experiment.) The proper handling of any hazardous chemicals is emphasized by caution statements throughout the manual. None of the experiments require special or elaborate equipment or chemicals. Nevertheless, the experiments do illustrate the fundamental ideas of chemistry and important laboratory techniques.

The first few pages of this manual include safety rules and a laboratory equipment list. In Experiment 1, students learn basic laboratory methods and measurement techniques. Experiment 2 provides experience with the idea of density. A variety of elements and compounds are investigated in Experiment 3. This is followed by Experiment 4 in which the differences between mixtures and solutions are observed along with some changes of state. Experiment 5 is the quantitative determination of the empirical formula of a compound.

In Experiment 7 the students determine the percent by mass of water in popcorn (a food) and in Experiment 8 they determine the percent by mass of water in a hydrate (a compound). Experiment 9 is designed to allow the students to analyze a vinegar solution and serves as a prelude to the concepts of stoichiometry. Methods used for the detection of some common ions and the detection of ions in common chemicals is the subject of Experiment 10. Experiment 12 is a study of the chemistry of oxygen and provides for the observation of several typical reactions. The observations of energy changes that accompany chemical reactions and a changes of state are included in Experiment 13, "Energy in Chemistry." The principles of stoichiometry are illustrated in concrete terms in Experiment 14.

Experiment 16 and Experiment 17 provide experience with the use and applications of gas laws. Experiment 19 is a study of solutions and solubility. "Conductivity of Solutions," Experiment 20, provides a demonstration of the conductivities of solutions of various electrolytes and nonelectrolytes. It is a good exercise to emphasize the nature of ionic solutions. Experiment 21 is an introduction to chemical equilibrium and Le Chatelier's Principle.

Precipitation reactions and the writing of net-ionic equations is the subject of Experiment 22. Experiment 23 provides an introduction to acid-base theory through the observation of acid-base reactions and the deductions of acid-base equations. In Experiment 24, the student learns to use the buret and how to titrate. A standard solution of sodium hydroxide is used to analyze acid solutions. The observation of redox reactions and the deduction of redox equations is covered in Experiment 25. Aspirin is synthesized in Experiment 26.

Four worksheets with practice problems are available for use in problem sessions or as nonlaboratory assignments. These are Exercise 6 (Formulas), Exercise 11 (Nomenclature), Exercise 15 (Stoichiometry), and Exercise 18 (Gas Laws).

The appendices include a table of the vapor pressures of water, a table of common laboratory acids and bases and an introduction to graphing. The introduction to graphing appendix includes an example of graphing along with some practice graphing assignments. One of these assignments includes the collection and graphing of experimental data.

I would like to thank all of the students and instructors who helped me prepare the fourth edition of this manual. Special thanks to Pat Blanchette for her outstanding typing.

T. R. Dickson

Aptos, California 1987

Contents

Laboratory Rules and Safety 1

Laboratory Desk Equipment 3

Typical Laboratory Equipment 4

1 Laboratory Techniques and Measurements 5

2 Density 21

3 Elements and Compounds 31

4 Chemicals, Mixtures and Solutions 39

5 Empirical Formula 49

6 Formula Worksheet 57

7 Popcorn: Water in a Mixture 59

8 Hydrates: Water in a Complex Compound 65

9 Analysis of Vinegar 73

10 The Detection of Common Ions 81

11 Nomenclature Worksheet 91

12 The Chemistry of Oxygen 93

13 Energy in Chemistry 103

14 Stoichiometry 113

15 Stoichiometry Worksheet 121

16 Gas Laws 123

17 The Molar Mass of a Gas 135

18 Gas Laws Worksheet 143

19 Solubilities and Solutions 145

20 Conductivities of Solutions 157

21 Equilibrium and Le Chatelier's Principle 165

22 Precipitation Reactions 175

23 Acids and Bases 183

24 Acid-Base Titration 197

25 Oxidation-Reduction Reactions 207

26 The Preparation of Aspirin 221

Appendix 1 The Vapor Pressure of Water 229

Appendix 2 Laboratory Acids and Bases 231

Appendix 3 Graphing 233

Laboratory Rules and Safety

1. Eye protection should be worn when you are working in the laboratory.

2. Follow the instructions given in the experiments. Never perform unauthorized experiments.

3. Read all labels before using chemicals. Make sure you have the correct chemicals.

4. When heating a liquid in a test tube or carrying out a reaction in a test tube, never point the mouth of the test tube at yourself or your neighbor.

5. Never taste a chemical. Avoid breathing gases given off in chemical reactions.

6. If dangerous gases are given off during an experiment, do the experiment under the hood as instructed.

7. Report any injury to the instructor at once, no matter how slight it may appear to be.

8. Never pour water into concentrated acid. Always slowly pour the acid into the water while mixing.

9. Return all waste chemicals to the containers indicated by your instructor. If no containers are available, pour all waste liquids directly down the drain and then run water down the drain. All waste solids (except those indicated) should be placed in the waste baskets. Do not throw solids in the sink.

10. Always moisten or lubricate glass tubing when it is being connected to rubber tubing or inserted in rubber stoppers.

11. If you spill any solid chemicals, clean them up. If you spill any liquid chemicals, clean them up with a towel. If you spill any acids or bases on the desk or floor, sprinkle some sodium hydrogen carbonate on the spill to neutralize it, then wipe it up.

12. If you splash any acid or base in your eyes, face, or hands, flush with plenty of water. Keep in mind the nearest source of water so that you can go to it as quickly as possible. If such a splash occurs, notify the instructor who will recommend any further action.

13. No eating, drinking, or smoking in the laboratory.

Laboratory Desk Equipment

2 Beakers, 100 mL

1 Beaker, 250 mL

1 Beaker, 400 mL

1 Crucible, porcelain

1 Crucible cover

1 Graduated cylinder, 10 mL

1 Graduated cylinder, 50 mL

1 Evaporating dish

2 Erlenmeyer flasks, 50 mL

1 Erlenmeyer flask, 125 mL

2 Erlenmeyer flasks, 250 mL

1 Funnel

1 Test tube rack

1 Thermometer, 110 ^{0}C

1 Crucible tongs

1 Tweezers

1 Test tube holder

1 Test tube brush

6 Test tubes, 15 cm

6 Test tubes, 10 cm

2 Watch glasses

1 Clay triangle

1 Wire gauze

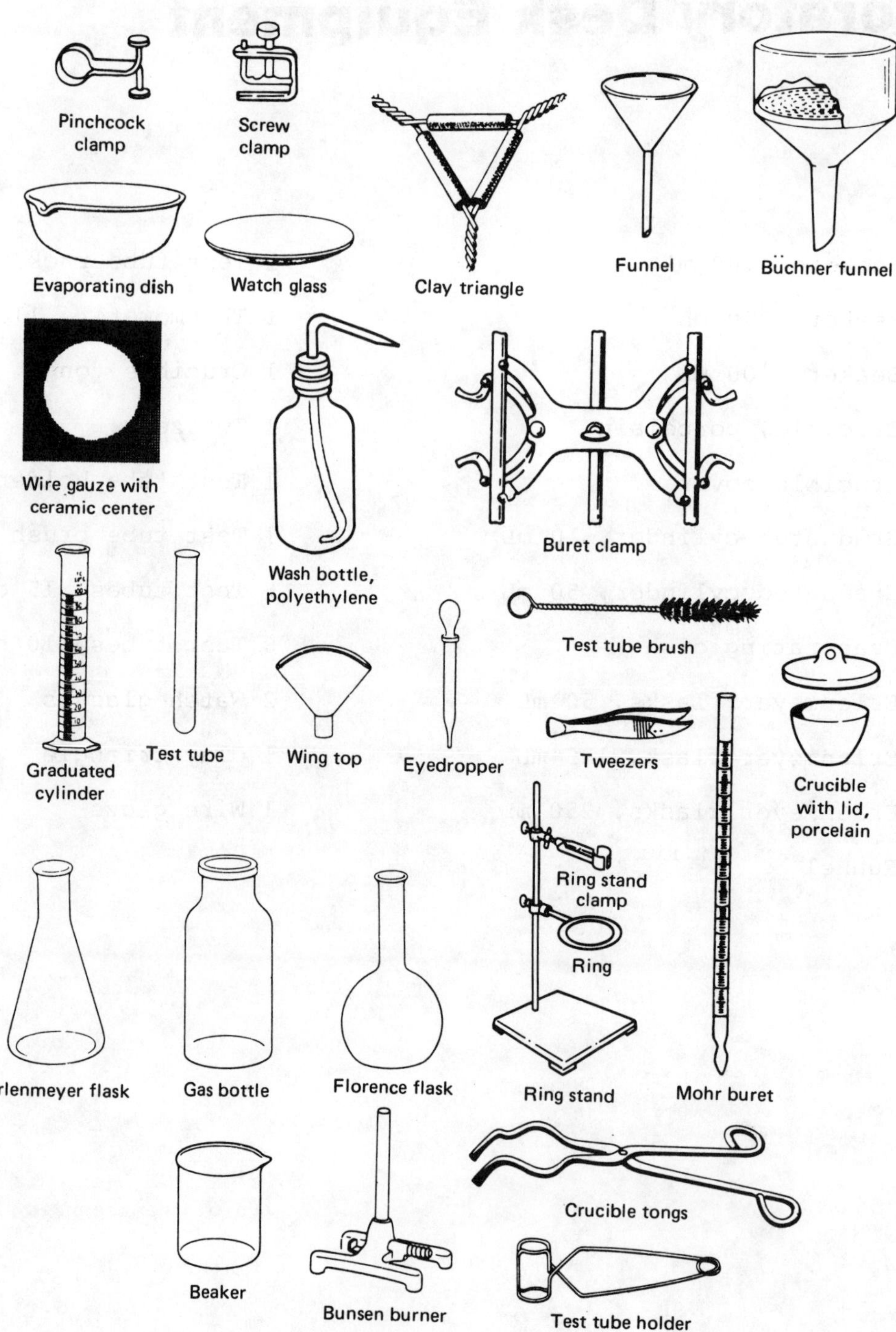

Typical Laboratory Equipment

1 Laboratory Techniques and Measurements

Objective

In this exercise you will learn of and practice several important laboratory techniques. These techniques will be used throughout this laboratory manual in a variety of experiments.

Discussion

Since chemistry is an experimental science, observations and making measurements are very important. Special measuring devices are used to measure the properties of objects. These measuring devices or instruments are typically calibrated to read in the appropriate metric unit. For example, length measurements are made using a calibrated ruler called a meter stick. Mass measurements are carried out using a balance which compares the mass of an object to calibrated masses called weights. Temperature measurement is made through the use of a thermometer.

A measuring instrument is calibrated to be read to a certain number of digits. Usually, a measuring instrument should be read to as many digits as possible. This sometimes requires estimating between the smallest calibrated divisions on the instrument. Such an estimation process is called interpolation and is described in Figure 1-1. Whenever you make a measurement, it is very important to always read the instrument as carefully as possible and to read the proper number of digits. In this exercise you will be instructed to make measurements to specific numbers of digits. When you make a measurement always record it as both a number and a unit (e.g., 2.32 cm, 5.78 g, 25.0 mL).

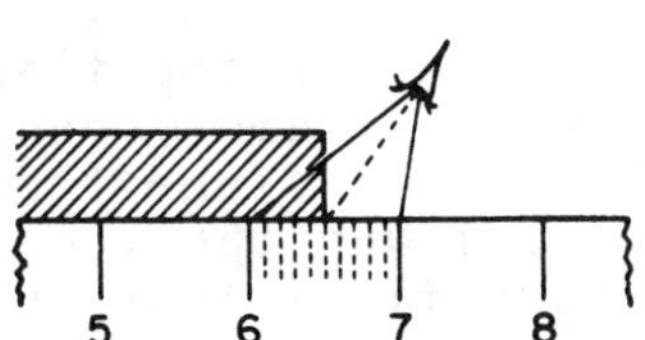

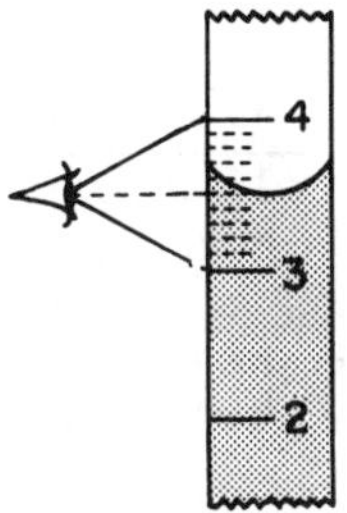

Fig. 1-1 Interpolation. The distance between divisions is visually divided into 10 equal portions and the visualized line closest to the object edge, marker, or bottom of the meniscus is noted. The distance between divisions may be divided into fewer parts depending on the calibration of the measuring device.

1. The Burner

A Bunsen burner mixes natural gas with air so that the gas can be burned and serve as a source of heat for laboratory experiments. Typical Bunsen burners are shown in Figure 1-2. The gas and air are mixed in the barrel of the burner. Note that there is an adjustable air intake vent on the barrel. The flow of gas is regulated by the gas valve connected to the gas pipe or, on some burners, there is a gas-regulating valve connected to the burner. To light the burner, turn on the gas valve to about half pressure (not full pressure). Hold a lighted match at the edge of the top of the barrel but not directly over the top of the barrel. If a flint striker is used, hold the striker over the top of the barrel and strike it.

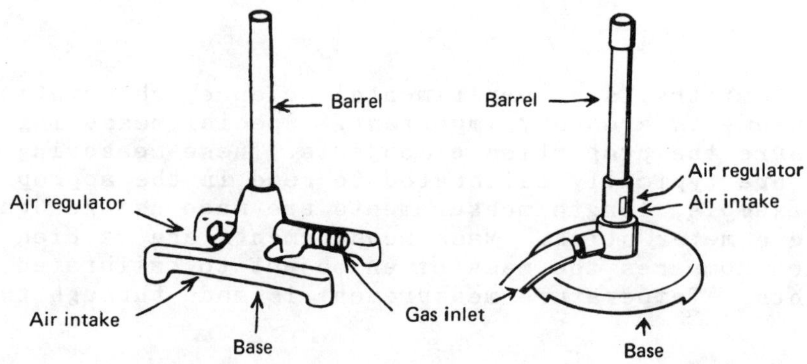

Fig. 1-2 Typical Bunsen Burners.

Once the burner has been lit, the flame can be regulated by adjusting the gas flow and the air intake vent. The flame on a properly adjusted burner will have the appearance of a pale blue inner cone and a pale violet outer cone. The hottest part of the flame is at the tip of the inner blue cone. If there is too much gas in the flame, it will have a yellowish appearance. To obtain a proper flame, open the air intake. Too much air and gas pressure will form a flame that separates from the burner top. To adjust such a flame, lower the gas pressure and decrease the amount of air. If you want a flame that is cooler than the normal flame, decrease the amount of air by adjusting the air intake. A low burning flame can be obtained by adjusting the gas pressure using the gas valve.

EXERCISE: Light your Bunsen burner and adjust the flame to give a pale blue inner cone and a pale violet outer cone. Sketch a side view of the burner showing the flame and indicate the colors of the cones. (A) Any letter in parentheses in a laboratory procedure refers to the place information should be recorded on the laboratory report sheet found at the end of the experiment.

---

2. DISPENSING CHEMICALS

When pouring liquid chemicals from glass-stoppered reagent bottles, remove the stopper and hold it in your fingers while carefully pouring the liquid into the desired container. This technique is illustrated in Fig. 1-3. When pouring from a screw cap bottle, set the cap down on the top so that it does not become contaminated. Be sure to put the correct cap on the bottle after you have used it. If you spill any liquid or drip some on the side of the bottle, clean it up. Never pour excess chemicals back into the reagent bottles since this can contaminate the reagents.

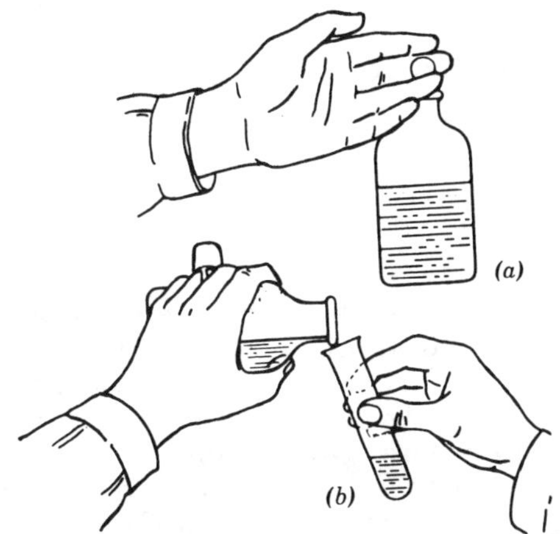

Fig. 1-3 Proper technique for pouring liquid from a glass-stoppered bottle.

When obtaining samples of powdered or crystalline solids from a jar, pour the desired amount of the solid onto a small piece of clean paper or into a clean, dry beaker. To pour a solid, do not turn the jar upside down and dump. Carefully tilt the jar and rotate it back and forth to work the solid up to the lip. Then, using the same back and forth rotation, allow the desired amount of solid to fall from the jar. This is illustrated in Fig. 1-4. Be careful when transferring a solid from a jar. If you take too much, leave the excess on a piece of paper for other students or throw the excess away. Never put any solid back into the jar. Furthermore, never put wooden splints, spatulas, or paper into a jar of solid unless your instructor indicates that this is permissible. Solids may be transferred by using a piece of paper that has been creased down the center. (See Fig. 1-4.)

EXERCISE: Take a large test tube to the appropriate station and pour a sample of liquid water into the tube from a glass-stoppered bottle. Place the tube in your test tube rack for use in the next section.

Obtain a square of paper and crease it down the center. Take the paper to the designated shelf or station and obtain a sample of solid salt by dispensing from a jar. Dispense a sample about the size of a pencil eraser. Keep the solid for the next section.

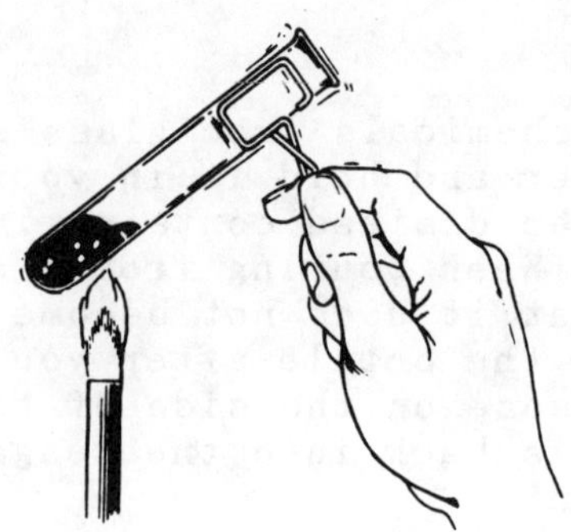

Fig. 1-6 Proper technique for heating a test tube of liquid.

3. HEATING CHEMICALS

Solid and liquid chemicals must be heated with care to prevent explosions and accidents. To heat liquids in beakers or flasks, the container can be placed on a wire gauze supported on an iron ring attached to a ring stand. As shown in Fig. 1-5 on the next page the burner can be placed under the gauze and the ring can be adjusted for efficient heating. When heating a liquid in a test tube, hold the tube with a test tube holder. The test tube holder prevents burned fingers. The test tube should be carefully heated by moving it back and forth in the flame so that the contents are evenly heated. (See Fig. 1-6 on the next page.) DANGER: Never hold the tube in the flame without moving it back and forth. As soon as the tube is heated, remove it from the flame. Failure to do this may cause the liquid to suddenly boil and fly out of the tube. Since this can be very dangerous, never point the mouth of the tube at yourself or your neighbor (or the instructor).

EXERCISE: Carefully pour the solid on the creased paper into the liquid in the test tube. Light your Bunsen burner and heat the contents of the tube in the flame using a test tube holder. Heat the tube for a while but do not let the liquid boil. If it begins to boil, remove from the flame, allow to cool slightly and continue heating. Describe what happens as the contents of the tube are heated.(B)

---

4. LENGTH MEASUREMENT

Examine a meter stick or metric ruler and note the parts into which it is subdivided. To what fraction of a meter can a length be measured with a meter stick?(C) Measure the length of a page in your laboratory book first in units of inches and then in units of centimeters.

Length of page 11.0 in. 28 cm (D) Express the inch measurement as a decimal fraction (i.e., 11-1/2 in. = 11.5 in.) ______ in. (E) Using these two measurements, calculate the ratio of centimeters to inches by dividing one by the other and express the answer to three digits (F). .392

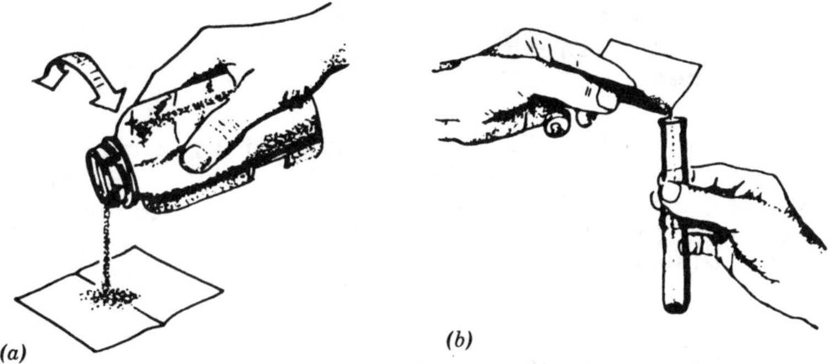

Fig. 1-4 Technique for transferring a solid. (a) Slowly rotate the jar until the desired amount of solid falls out. (b) Using a creased paper, pour the solid into the test tube.

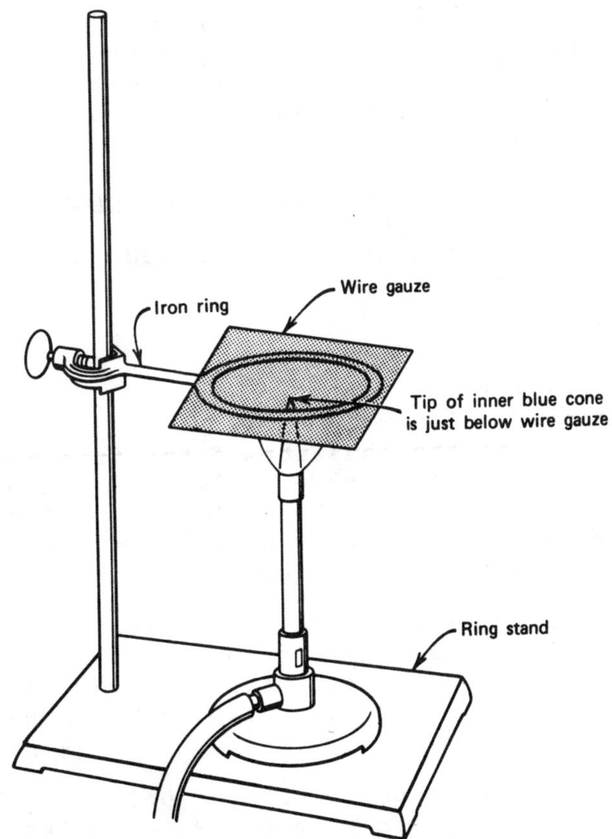

Fig. 1-5 Apparatus for heating liquids. The flask of beaker can be placed on the gauze for heating purposes.

5. TEMPERATURE MEASUREMENT

Examine one of the Celsius thermometers on display and read it to the nearest 0.1 °C. Record the temperature of the room (G). NOTE: If you break a mercury thermometer, notify your instructor immediately so that the mercury can be properly cleaned up.

6. VOLUME MEASUREMENT

(a) Examine your 10-mL graduated cylinder. Note the large and small calibration lines on the cylinder. What volume can be contained between adjacent lines in the cylinder? (It is possible to read a volume to the nearest 0.1 mL in this cylinder?) (H)

(b) Three partially filled graduated cylinders are on display. For each of these cylinders indicate the amount of volume contained between adjacent lines. For instance, the volume between lines may be 10 mL or 1 mL or a fraction of a milliliter. Read the volume of the liquid in each cylinder as indicated below.

Cylinder	Volume between lines	Volume of liquid in cylinder	
1000-mL	10 mL	640 mL	(to nearest 1 mL) (I)
100-mL	1 mL	75 mL	(to nearest 0.5 mL)(J)
10-mL	.2 mL	5.8 mL	(to nearest 0.1 mL)(K)

(c) Fill a small test tube with water and pour the water into your 10-mL graduated cylinder to measure the volume of the tube. Record the volume.

Volume of tube 10 mL (L)

How full would you have to fill this tube so that you would have about 2 mL of water? (M)

7. BALANCES AND WEIGHING

The most common types of balances found in the laboratory are the centigram balance and the top loader balance. A typical centigram balance is shown in Fig. 1-7 on the next page. These types of balances are usually designed to allow objects to be weighed to the nearest 0.01 g. That is, to the nearest centigram.

USE OF THE CENTIGRAM BALANCE There are various models and brands of centigram balances. The description that follows refers to a magnetically dampened model. Your instructor will explain how the models in your laboratory may differ from this type. The steps involved in using the balance are listed below.

1. Determine the rest point or zero point of the balance by setting the sliding weights to zero and allowing the beam to swing freely. When the movement of the beam stops, note the position of the pointer on the pointer scale. The pointer should come to rest so that it points at or near the zero on the pointer scale. This is the rest

point of the pointer. If the pointer is not near the zero on the pointer scale, ask your instructor to adjust the balance.

2. Place the object to be weighed on the pan of the balance. The beam will tip so that the pointer goes up the scale. Move the large 10 g weight until the beam tips and the pointer goes down the scale. Then move the 10 g weight back one notch. Repeat with the 1 g weight. Once the proper settings of these weights have been found, the sliding 0.01 g weight is moved to the right to a position in which the beam swings freely. Allow the pointer to come to rest. At this time, the position of the pointer on the pointer scale is noted to see if it corresponds to the rest point. If it does not, the sliding mass is moved slightly to the right or left until the position of the pointer corresponds to the original rest point. Carefully read and record the mass of the object by noting all of the weight settings. It is always a good idea to double check the reading of the mass.

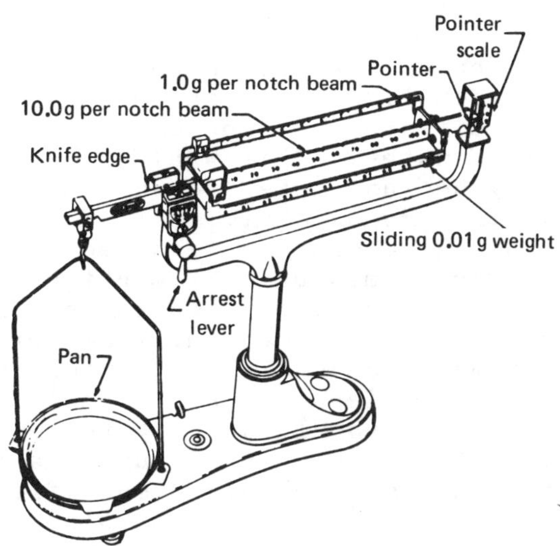

Fig. 1-7 A typical laboratory balance.

USE OF THE TOP LOADER BALANCE

Top loader balances come in a variety of designs. Your instructor will demonstrate the use of this type of balance if one is available for your use.

WEIGHING SAMPLES

The general rules for using balances are:

1. Never place chemicals directly on the balance pan. Always use weighing paper or some container.

2. If you have trouble using a balance, contact your instructor.

3. Always clean up any chemicals that you spill on the balance or in the weighing area.

When a sample of a substance is weighed on a piece of paper or in a container, the mass of the paper or empty container is first determined. Then the substance is placed on the paper or in the container and the total mass is determined. The mass of the substance is then found by subtracting the mass of the paper or container from the total mass. This is called weighing by difference.

EXERCISE: Obtain a sample of a substance to be weighed from your instructor of the laboratory stockroom. First weigh a clean beaker to the nearest 0.01 g.(N) Place the object to be weighed in the beaker and weigh the beaker plus the object to the nearest 0.01 g.(O) Now subtract the smaller mass from the larger mass to find the mass of the object.(P)

---

8. ROUGH AND PRECISE MEASUREMENTS

In this manual, you will sometimes be instructed to measure out a definite quantity and sometimes only an approximate quantity. It is important to note the difference. For example, you may be instructed to obtain about 2 mL of water. This would be an approximate quantity. At other times you may be instructed to obtain about 2 mL of water to the nearest 0.1 mL. This means that, while the volume should be somewhere around 2 mL, its value must be recorded to the nearest 0.1 mL. The results of this type of measurement might be 1.8 mL, 2.2 mL, or 2.3 mL. Occasionally, you may be instructed to measure 2.0 mL of water. In such a case, you need to very carefully measure out the required amount. Since such a measurement is quite time consuming, never measure out exact amounts unless you are directed to do so. However, this does not mean that you should not try to read your measuring instrument as closely as possible when you do make a measurement.

When you are instructed to obtain a specific volume of a liquid, it must be precisely measured out using a volume measuring device such as a graduated cylinder. When you are instructed to obtain an approximately volume (i.e., about 2 mL), you may pour into your graduated cylinder until you have approximately the amount you want.

When you are instructed to obtain an approximate mass (i.e., obtain about 1 g) of a substance, this can be done by weighing a piece of paper (or beaker) and then setting the balance to the weight of the paper plus the weight of the sample desired. Then, the substance can be slowly poured onto the paper until movement of the balance beam indicates that the approximate mass has been taken. When you are instructed to weigh a sample of material precisley (i.e., obtain a sample of about 1 g weighed to the nearest 0.01 g), you must obtain the sample and weigh it as carefully as possible.

9. GLASS WORKING

When setting up apparatus for experiments, it is sometimes necessary to use glass tubing for glass connecting tubes or for such things as stirring rods. It is important to learn how to handle glass tubing. Always wear eye protection when cutting glass.

CUTTING GLASS TUBING Obtain a length of glass tubing. Using a triangular file, a glass knife or a glass cutting wheel, make a scratch on the tubing at the desired point. Make a small scratch and do not extend it all the way around the tubing. If using a file, make the scratch with an outward motion of the file. Do not saw the tubing with the file. Once scratched, pick the tubing up and hold it so that the thumbs are together on the opposite side of the scratch, and then gently push the glass away from you with your thumbs. This is shown in Fig. 1-8. Wear eye protection when cutting glass.

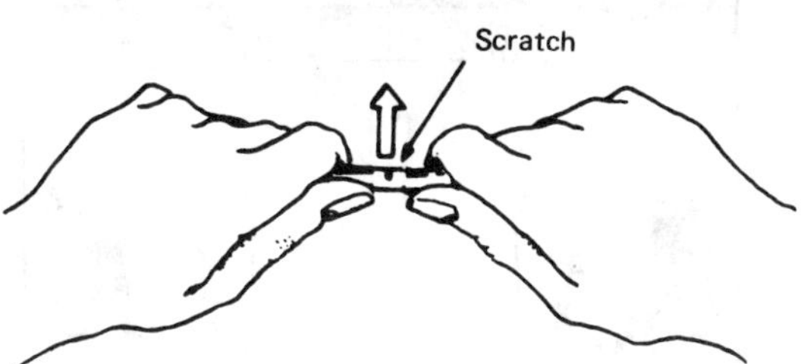

Fig. 1-8 Cutting glass tubing.

FIRE POLISHING GLASS Freshly cut glass has very sharp edges that must be smoothed in the Bunsen flame. Always fire polish glass tubing. Fire polishing is done by rotating the end of the glass tubing in the Bunsen flame until the sharp edges are smooth. If the tube is held in the flame too long, it will seal. After polishing, a tube should be placed on a ceramic sheet or wire gauze to cool. Be careful of hot tubing. Do not touch the tubing until it is cool. Fire polish all ends of glass tubing.

SEALING TUBES AND MAKING STIRRING RODS A stirring rod can be constructed from glass tubing or from solid glass rod. Glass rod makes the best stirring rods. If rod is used, cut the desired length and polish each end. If tubing is used, cut a piece about 5 inches longer than the desired length of the rod. Hold the tube with both hands and place it at the tip of the Bunsen flame about 2 inches from one end of the tubing. Rotate or roll the tube in your fingers as it heats. When the tubing begins to melt at the point of heating, remove it from the flame and gently pull both ends to stretch the tubing. As the tubing stretches it will decrease in diameter. Set the tube aside to cool.

After cooling, use a file to cut the tube at the constriction. Place the constricted end of the tube in the hot part of the Bunsen flame and rotate the tube until the end melts and seals. Make sure to heat long enough to seal any small holes in the tip. After cooling, the other end of the tube can be sealed in a similar fashion.

BENDING GLASS TUBING Glass tubing can be bent to desired angles for use in glass delivery tubes and glass connecting tubes for use in experiments. When bending glass, always start with a length of glass that is longer than needed. This allows for easy heating without burning your fingers. After the bent glass has cooled, it can be cut to the desired length and fire polished. Attach a wing top flame spreader to the Bunsen burner and light the burner. Hold the tubing in the flame where the bend is desired and slowly rotate the tubing to heat evenly. (See Fig. 1-9.) It is important to continuously rotate the tubing while heating and to heat it over the entire section to be bent. When the glass is soft and begins to droop, remove it from the flame and bend the two ends in an upward direction to form the desired angle (see Fig. 1-9) and then set it on the ceramic sheet or wire gauze to cool.

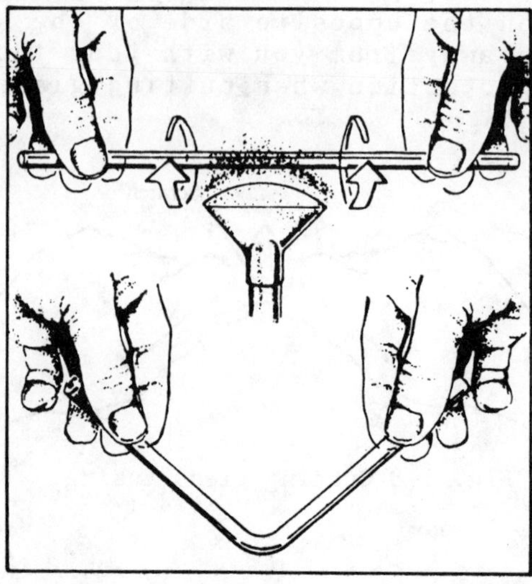

Fig. 1-9 Bending glass tubing.

INSERTING GLASS TUBING IN RUBBER STOPPERS The insertion of glass tubing into rubber stoppers can be dangerous. First of all, make sure that all tubing is fire polished and cool. Lubricate the end of the tubing and the edge of the hold in the stopper with water or glycerol. Your hands can be protected with gloves or with a towel if desired. Place the tubing in the hole and work it in by twisting in the other direction. Never push the tubing into the hole. Always twist the tubing by holding the portion of it that is closest to the stopper. Always be careful and do not hurry. If you have trouble, tell your instructor. If the tubing breaks and you cut yourself, tell your instructor immediately.

EXERCISE: Obtain some glass tubing and make the glass objects shown in Fig. 1-10 on the next page. Note that the glass items in the figure are not shown to scale so you will need a metric rule to estimate lengths. For the angular and short tubes work with much longer lengths of tubing than you need and then cut the tubing to the desired lengths after bending or sealing. Do not forget to fire polish after cutting tubing. Furthermore, be careful not to touch the hot glass. After making your glass objects, obtain your instructor's approval on your report sheet. (Q)

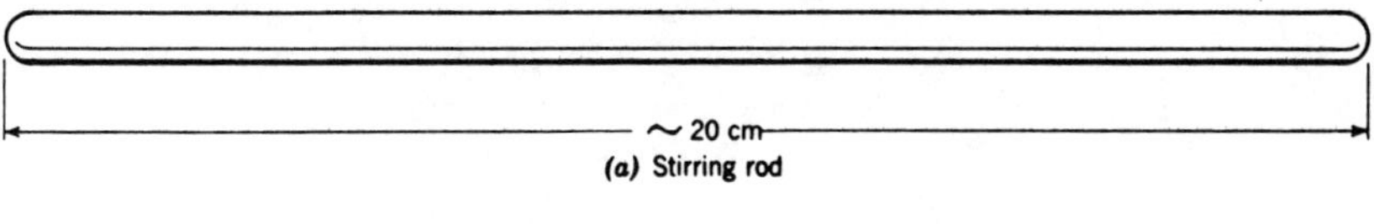

(a) Stirring rod

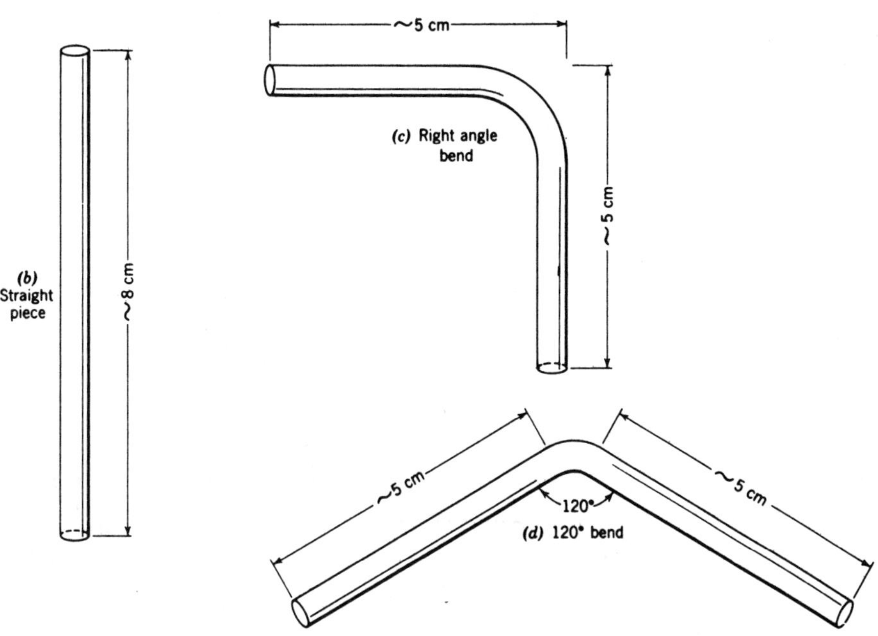

(b) Straight piece
(c) Right angle bend
(d) 120° bend

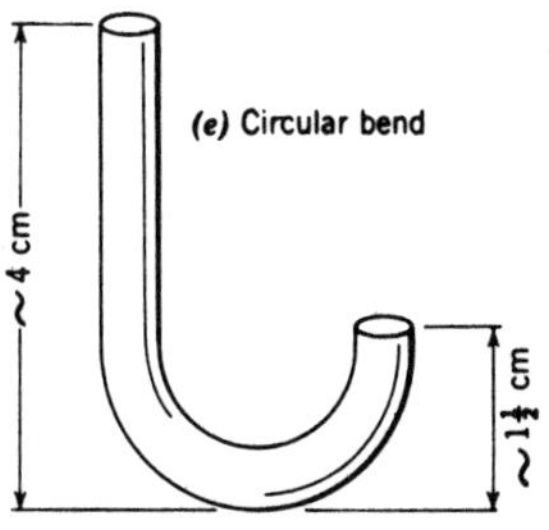

(e) Circular bend

Fig. 1-10

2 Density

Objective

The purpose of this experiment is to investigate the topic of density by determining the densities of some chemicals.

Discussion

There is a lesson in the old bromide: "Which is heavier, a pound of feathers or a pound of lead?" Neither is heavier since a pound of anything is still a pound. It is not possible to compare masses of objects unless it is done on the basis of the same portion of each object. For instance, it is possible to say that one person is heavier than another person since the comparison is on a pound per person basis. Since all chemicals have mass and occupy space, it is possible to compare chemicals by stating the amount of mass contained in a specific volume of each chemical. Density is defined as the mass per unit volume of a chemical where mass is usually expressed in grams and volume in cubic centimeters or milliliters. Density can be determined by measuring the mass and volume of a sample of a chemical and then dividing the mass by the volume:

$$\text{Density} = \frac{\text{mass}}{\text{volume}} \quad \text{or} \quad D = \frac{m}{V}$$

The density of any chemical is a characteristic property of the chemical and can help identify the chemical. Furthermore, density can be used to relate the mass of a sample of the chemical to the volume or the volume to the mass. That is, the density can be used as a property conversion factor to find the volume given the mass,

$$V = m\left(\frac{1}{D}\right)$$

or to find the mass given the volume,

$$m = VD$$

In this experiment, the densities of several chemicals are determined and densities are used as conversion factors.

The volumes of liquids are often determined in the laboratory by use of graduated cylinders. These cylinders are calibrated in the factory so that they will contain certain specific volumes. Lines are etched on the outside surface of the cylinder to indicate the positions corresponding to various volumes. Distances between subsequent lines on the cylinder correspond to specific volumes. Volumes of liquids can be measured by pouring a liquid sample into the cylinder and observing the position of the surface of the liquid. Some liquids, such as water, do not form a flat surface when placed in the cylinder, but instead they form a concave surface called a meniscus. Consequently, graduated cylinders are calibrated so that the volumes can be read by observing the lowest portion of the meniscus; the bottom of the meniscus. The volume of the liquid can be measured by observing the position of the meniscus with respect to the calibration lines on the cylinder. Consider an example of the reading of the volume of a liquid in a graduated cylinder. In Fig. 2-1b, the position of meniscus coincides with the 15-mL mark and the volume

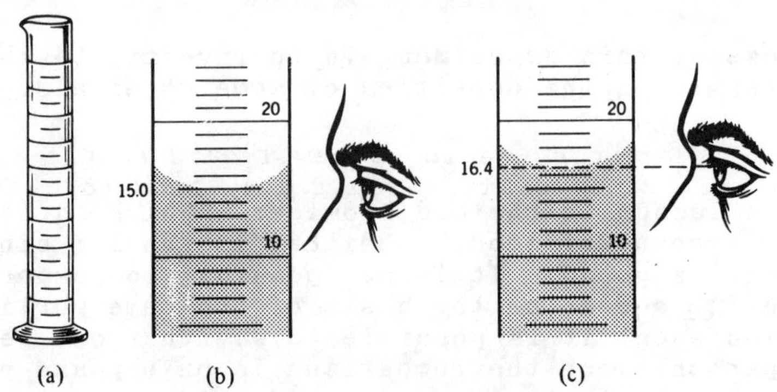

Fig. 2-1

of the liquid would be read as 15.0 mL. However, in Fig. 2-1c the meniscus level does not coincide with any particular line but falls between lines. The estimation of the position of the meniscus in this case illustrates a very important method involved in the measurement: interpolation. Interpolation can be carried out by visually splitting the distance between two lines into equal parts and then approximating the position at which the meniscus lies between the two lines. In the figure, if the distance between lines is split into 10 parts then the meniscus appears to lie at the 16.4-mL position. The interpolation method is very important in volume measurements. The number of digits to be read by interpolation depends on how the volume measuring device is calibrated.

Laboratory Procedure

The measurement of the masses of objects is to be carried out to the nearest 0.01 g. Be careful to use the balances correctly and to use the same balance for subsequent weighings. The volumes of chemicals will be determined using graduated cylinders. Since the number of significant digits in the volumes will dictate the number of digits in your calculated densities, be sure to read the volumes carefully. When you make experimental measurements it is desirable to make more than one set of measurements as time permits. To provide for more reliable results, the densities of the chemicals in this experiment should be measured twice and an average determined. The idea is to make measurements for the density of a chemical using a sample. Then, the measurements are repeated on a new sample. The two experimental densities can be averaged.

1. DENSITY OF WATER

Dry your 10-mL graduated cylinder and find its weight to the nearest 0.01 g.(B) Pour about 9 mL of water into the cylinder and read the volume to the nearest 0.1 mL.(D) Now weigh the cylinder plus the water to the nearest 0.01 g.(A) Calculate the mass of the water sample (C) and the density of the water.(E) Dry the cylinder and repeat the above for a new water sample. Finally, find the average of your two densities by adding them together and dividing by 2.(F)

2. DENSITY OF AN UNKNOWN LIQUID

Thoroughly dry your 10-mL graduated cylinder and weigh it to the nearest 0.01 g.(H) Pour about 9 mL of unknown liquid into the cylinder and read the volume to the nearest 0.1 mL.(J) Now weigh the cylinder plus the liquid to the nearest 0.01 g.(G) Calculate the mass of the liquid sample (I) and the density of the liquid.(K) Wash and dry the cylinder and repeat the above for a new liquid sample. Finally, find the average of your two densities by adding them together and dividing by 2.(L)

3. DENSITY OF SOLIDS

The density of an irregular solid can be found by first determining the mass of a sample and then placing the sample in a graduated cylinder partly filled with water (or some liquid in which the solid does not float). The solid will displace a volume of liquid equal to its volume. Thus, by noting the position of the meniscus before and after the addition of the solid, the volume of the solid can be deduced. (See Fig. 2-2.)

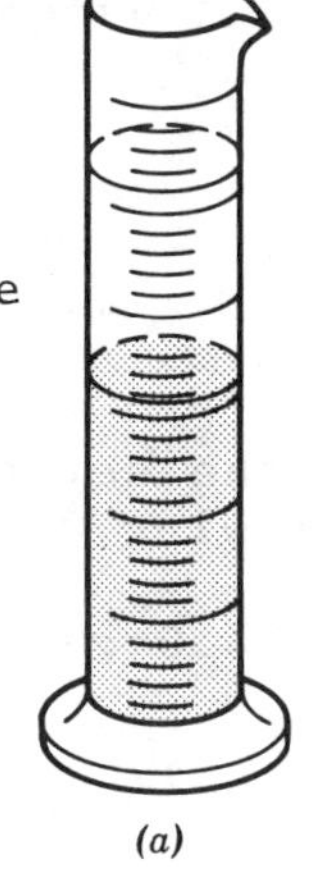

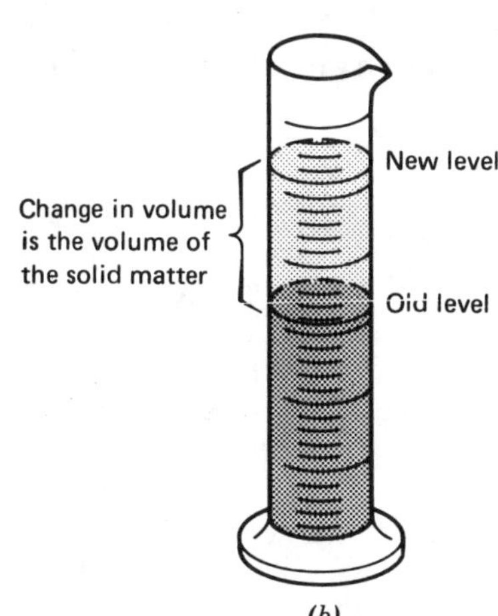

Fig. 2-2 Determination of the volume of a solid.

A. DENSITY OF SULFUR Obtain a few pieces of dry sulfur. Weigh a piece of weighing paper to the nearest 0.01 g. Place the sulfur on the paper and determine the mass to the nearest 0.02 g. By subtraction, find the mass of the sulfur. Repeat the process for another sample of sulfur. Mark the papers to prevent mixing up the samples.

	1	2	
Mass paper and sulfur	----------	----------	
Mass paper	----------	----------	
Mass sulfur	----------	----------	(M)

Fill your 10-mL graduated cylinder with about 5 mL of water and read the volume to the nearest 0.1 mL.(O) Carefully place the previously weighed sulfur lumps into the cylinder. Agitate slightly to be sure that no air bubbles are trapped. Read the new position of the meniscus to the nearest 0.1 mL.(N) (Return the damp sulfur lumps to the proper container.) Calculate the volume of the sulfur (P) sample and the density.(Q) Repeat the process for the second sulfur sample and find the average density.(R)

B. DENSITY OF UNKNOWN SOLID Obtain two samples of an unknown solid. Find the mass of each sample and the volume by water displacement as described above for sulfur. Calculate the average density.

Mass solid and paper or container	__________	__________	
Mass paper or container	__________	__________	
Mass solid	__________	__________	(S)
Volume water plus solid	__________	__________	(T)
Volume water	__________	__________	(U)
Volume solid	__________	__________	(V)
Density solid	__________	__________	(W)

Average density (X)

4. DENSITY AS A FACTOR

On the report sheet give the setup of the calculation and the calculated answer for the following.

(a) What volume would a 35.0 g sample of the unknown liquid in part 2 have? (Y)

(b) What mass would a 125 cm^3 sample of the unknown solid in part 3 have? (Z)

5. THICKNESS OF METAL FOIL

To directly measure the thickness of a sample of aluminum foil a special instrument is needed. However, the thickness can be estimated indirectly by viewing a sheet of foil as a thin rectangular solid having a length, width and height. The height is the thickness. Since density relates mass to volume the volume of a rectangular sample can be found from its mass. If we know the length, width and volume of a rectangular solid it is possible to calculate the height.

Obtain a sheet of aluminum foil. Smooth the sheet on your laboratory book and, using a meter stick, carefully measure the length of the edge of the sheet to the nearest 0.1 cm.(BB) If the sheet is not square, measure the length of adjacent edges. Use the balance to find the mass of the sheet.(AA) Find the area, A, of the sheet from the length of the edge or edges. The area is the product of the length and the width or the square of the length of the edge if the sheet is a square.

Calculate the approximate thickness of the foil as follows. Assuming that the foil is a rectangular solid, the volume is given by the area of the face times the height (thickness). This can be expressed as

$$V = At$$

where V is the volume, A is the area, and t is the thickness. If the density of aluminum is 2.70 g/cm^3, the volume of the foil sample can be determined from the mass and the density. Once the volume is known, it can be used along with the area to find the thickness of the foil. Calculate the thickness of the aluminum foil in centimeters. Express the thickness in terms of millimeters.(CC)

6. THICKNESS OF A COIN

The thickness of a coin can be indirectly estimated if we assume that a coin is a cylindrical solid. For such a solid the volume is given by the product of the area of the circular base and the height or thickness. The area of the base can be found by measuring the diameter of the coin. The volume of the coin can be found using the mass of the coin and the density of the coin which is known. Then the thickness of the coin can be calculated using the area of the circular base and the volume.

Measure the diameter of a nickel or quarter to a 0.1 cm.(EE) Now determine the mass of the coin.(DD) Estimate the thickness of the coin as follows. Assuming that the coin is a cylinder, the volume is calculated by the formula

$$V = \pi (d/2)^2 t$$

where V is the volume, π is 3.14, d is the diameter, and t is the thickness (height). Nickels, dimes and quarters are composed of nickel and copper metals. If the densities of the metals in the coin are known (Ni 8.9 g/cm^3 and Cu 8.9 g/cm^3), the volume of the coin is found from the mass. Once the volume is known, it can be used along with the diameter to find the thickness of the coin. Calculate the thickness of your coin.(FF)

Obtain a sheet of aluminum foil. Smooth it out on your laboratory bench and, using a meter stick, carefully measure the length of the edge of the sheet to the nearest 0.1 cm. (14) If the sheet is not square, measure the length of adjacent edges. Use the balance to find the mass of the sheet. (15) Find the area (A) of the sheet from the length of the edge (or edges). The area is the product of the length and the width or the square of the length of the edge if the sheet is a square.

Calculate the approximate thickness of the foil in millimeters. Assuming that the foil is a rectangular solid, the volume is equal to the area of the face times the height (thickness). This can be expressed as

where V is the volume, A is the area, and t is the thickness. Since the density of aluminum is 2.70 g/cm³, the volume of the foil can be determined from the mass and the density. Once the volume is known, it can be used along with the area to determine the thickness of the foil. Calculate the thickness of the aluminum foil in centimeters and then express the thickness in terms of millimeters. (20)

D. THICKNESS OF A COIN

The thickness of a coin can be indirectly estimated by assuming that a coin is a cylindrical solid. For such a solid the volume is given as the product of the area of the circular base and the height or thickness. The area of the base can be found by measuring the diameter of the coin. The volume of the coin can be found using the mass of the coin and the density of the coin [illegible]. Once the thickness [illegible] the coin can be calculated using the area of the circular base and the volume.

Measure the diameter of a nickel or quarter to 0.1 cm. (21) Then determine the area of the coin. (22) Estimate the thickness of the coin as follows. Assuming that the coin is a cylinder, the volume is calculated by the formula

$$V = \pi d^2/4 \; t$$

where d is the diameter, π is 3.14, [illegible] and t is the thickness (height). Nickels, dimes and quarters are [illegible] of nickel and copper metals. The density of the metal in the coins is known [illegible] 8.9 g/cm³. The volume of the coin is found from the mass. Once the volume is known, [illegible] together with the diameter, find the thickness of the coin. Calculate the thickness [illegible]

[illegible]

3 Elements and Compounds

Objective

To observe some elements and compounds, to isolate some elements from compounds, to form some compounds, and to form a gaseous compound.

Discussion

The objects and materials in our environment are forms of matter. Matter can be separated and classified into various categories. A pure chemical is a homogeneous form of matter that has the same properties throughout. There are two types of pure chemicals. Compounds are pure chemicals that can be separated into simpler chemicals called elements. Elements are pure chemicals that cannot be separated into simpler chemicals. Elements are elementary or fundamental forms of matter. A compound is composed of two or more chemically combined elements. Elements can be isolated from compounds by the use of special chemical processes. Compounds are formed through chemical combinations of elements.

In this exercise you will observe the appearances and natures of some common elements and compounds. The electrical conductivities of some elements will be tested. Pure chemicals can occur as solids, liquids or gases and in a variety of forms and colors. Two different elements will be isolated from compounds, two compounds will be made from elements. Finally, a colorless, gaseous compound will be formed and its properties investigated.

Laboratory Procedure

1. SOME COMMON ELEMENTS

Observe the samples of elements on display and fill in the table given on the Report Sheet. (A)

2. SOME COMMON COMPOUNDS

Observe the samples of compounds on display and fill in the table given on the Report Sheet. (B)

3. ELECTRICAL CONDUCTIVITIES OF SOME ELEMENTS

Several elements are located near a conductivity apparatus. Using this apparatus as directed by your instructor, test the conductivities of the samples. Your instructor may choose to demonstrate this part of the experiment. Record your observations in the table given on the Report Sheet. (C)

PRECAUTION: In the next sections you will be using several chemicals and you will be heating chemicals. Always wear safety goggles or glasses when working with chemicals. Take care in handling chemicals and hot objects. Hydrochloric acid and sulfuric acid should not be spilled or splashed. If you splash acid on your clothing or skin, wash it off with large amounts of water and inform your instructor.

4. FORMING ELEMENTS FROM COMPOUNDS

(a) Silicon from Silica Sand

In this exercise the element silicon will be formed from silica sand. Silica sand contains a silicon-oxygen compound. When silica sand is heated with magnesium, the magnesium displaces the silicon in the compound forming elemental silicon.

Place several pieces of magnesium turnings (about 0.2 g) in a clean, large test tube. Add enough silica sand to just cover the pieces of magnesium. Hold the tube with a test tube holder making sure that the holder is near the top of the tube. Hold the tube at an angle and place the bottom portion of the tube in the hot part of a Bunsen burner flame. Heat the tube in the flame for several minutes until you see the noticeable orange glow in the contents. After you see this glow in the contents, take the tube out of the flame and hold it while it cools for a few minutes. Place the tube in a test tube rack or a small beaker until it cools to room temperature. Do not touch the hot tube. Go to part (b) of this exercise and return to this part after finishing part (b).

Obtain a clean piece of paper and pour the cooled contents of the tube onto the paper. Observe the contents and describe the difference in appearance of the contents before and after heating. (D) Incidentally, the test tube used for this exercise will be permanently changed. Glass is also a compound of silicon and some silicon will be formed on the surface of the glass.

(b) Copper from a Copper-Oxygen Compound

In this exercise the element copper will be isolated from a copper-oxygen compound. When the compound is heated in the proper part of a flame and cooled quickly, copper metal will be formed. Copper can be recognized by its distinctive color.

Obtain a sample of copper-oxygen compound about the size of a pencil eraser. Place the sample on a piece of paper. Obtain a piece of copper metal in the form of a thin sheet. Place it on the piece of paper. Describe the appearance of the copper-oxygen compound and the elemental copper. (E)

Fill a small beaker with cold water. Light your Bunsen burner and adjust the flame so that you can see a blue inner cone. Pick up a sample of the copper-oxygen compound about the size of a pea using the tip of a stainless steel spatula. Carefully place the tip of the spatula in the Bunsen flame so that the sample is just above the blue inner cone. Heat the entire sample in this position for a minute or two making sure that the entire sample glows orange hot. Quickly immerse the tip of the spatula into the beaker of water. Remove the wet sample and place it on the piece of paper. Compare this sample to your sample of copper metal and describe the comparison. (F) Save the piece of copper sheet for part 5 of this experiment. Return to part (a) and finish your observations.

5. FORMING COMPOUNDS FROM ELEMENTS

(a) Copper-Sulfur Compound

Copper and sulfur will be heated to form a distinctive copper-sulfur compound. Obtain a small piece of copper metal in the form of a sheet. Place a sample of powdered sulfur about the size of a pea on a piece of paper. Hold the copper sheet by its edge using tweezers. Set up a Bunsen burner in the fume hood. Using a spatula, place about one half of the sulfur sample on top of the copper sheet. Hold the sulfur-covered copper in a Bunsen flame for a short time until no more sulfur can be seen. Remove the sheet from the flame, let it cool and observe it. Describe the difference between the elements copper and sulfur and their compound which has formed on top of the sheet. The sheet will also contain some remaining copper. (G)

(b) Aluminum-Chlorine Compound

Elemental aluminum will be mixed with some hydrochloric acid and heated to form an aluminum-chlorine compound.

Place a small piece of aluminum foil in a clean, dry test tube. Set up a Bunsen burner in a fume hood and continue this exercise in the hood. Add 4 drops of 6 M hydrochloric acid to the tube containing the aluminum. This is a special solution of hydrochloric acid and you find it in a bottle labeled 6 M hydrochloric acid or 6 M HCl. Handle this acid solution with care. Heat the tube strongly until the water boils off. If the contents are slightly gray or some aluminum remains, add 1 drop of hydrochloric acid and continue heating. Heat strongly until all of the water evaporates. Describe the differences in appearance of the aluminum and the aluminum-chlorine compound. (H)

6. A GASEOUS COMPOUND: CARBON DIOXIDE

Some compounds are gases and unless they have a color, it is not possible to see them. Such colorless gases can be detected by their chemical behavior. In this section, the gaseous compound carbon dioxide will be formed and detected. For this experiment you will need a 250-mL beaker, a 100-mL beaker and a few wooden splints. Light one of your splints in a Bunsen flame. The burning of the wood is supported by the oxygen in the air. Immerse the flaming splint into one of the beakers and note that, as expected, it continues to burn. Put out the splint.

Pour a sample of solid sodium bicarbonate, $NaHCO_3$, into the 250-mL beaker to cover the bottom of the beaker to a height of about 3 mm. Measure about 10 mL of dilute sulfuric acid into a graduated cylinder. This is a special solution of sulfuric acid acid and you find it in a bottle labeled 3 M sulfuric acid or 3 M H_2SO_4. Handle this acid solution with care. Pour the acid into the beaker containing the sodium bicarbonate. Note the vigorous bubbling as the carbon dioxide, CO_2, gas is formed. Since the gas is colorless, you will not be able to see it fill the beaker as it pushes the air aside. Carbon dioxide will not support the burning of wood. Light a wooden splint and immerse it into the gas in the beaker. Describe your observations. (I)

In time, the carbon dioxide will diffuse from the beaker into the surrounding air. But, before it does, we can observe an interesting property of the gas. Carbon dioxide has a greater density than air so it is possible to move it from one beaker to another by pouring it. Rinse out the 250-mL beaker with some water. Place a new sample of sodium bicarbonate into the beaker and pour in about 10 mL of dilute sulfuric acid (3 M H_2SO_4). Lift the beaker containing the carbon dioxide and tip it to pour the gas into the 100-mL beaker. Do not pour any of the liquid from the larger beaker but just tip it enough to pour the carbon dioxide gas. Light another wooden splint and immerse it into the 100-mL beaker. Light it again and again immerse it into the 100-mL beaker. Record your observations and explain your results. (J)

4 Chemicals, Mixtures and Solutions

Objective

The purposes of this exercise are to observe some common chemicals and solutions, to prepare a mixture and a solution, to carry out the filtration process, to determine the boiling point of a liquid and to observe melting, freezing, and sublimation.

Discussion

The materials and objects that make up the environment can be separated into a large variety of pure substances called chemicals. A pure chemical is a homogeneous form of matter with definite composition and is characterized by specific properties. A chemical may occur as a solid, liquid, or gas and it will have a specific color or lack of color.

Solids can occur in a variety of forms ranging from large crystals or chunks to powders. For instance, sugar can occur as large crystals (rock candy), be ground into small crystals (granulated sugar), or powdered as powdered sugar.

A pure chemical is homogeneous. Homogeneous means to be made up of the same parts or that all portions are the same and have the same properties. A pure chemical will have a fixed composition. That is, any two samples of a specific pure chemical will have the same composition as long as they are pure and contain insignificant amounts of impurities.

You might wonder if mixing sugar in water produced a compound since the resulting mixture appears to be pure and homogeneous. Mixtures of this kind do appear to be homogeneous but do not have definite composition. That is, sugar and water mixtures can contain variable amounts of sugar in water. Homogeneous mixtures of variable composition are called solutions. Solutions are mixtures of chemicals that dissolve in one another. Hold on, you may say, we can see the parts of a heterogeneous mixture, but how can a solution that looks pure be a mixture. First of all you have to realize that matter can be broken up into tiny pieces and mixed together. Some mixtures, like milk, appear to the eye to be pure chemicals but are actually heterogeneous mixtures. The difference between heterogeneous mixtures and solutions is that solutions involve very small particles of matter (far below even microscopic observation) that are intimately mixed, and heterogeneous mixtures involve larger aggregates of such particles. Solutions are quite commonly used in the laboratory.

Solutions that are made by dissolving one chemical in another are called binary solutions. In a binary solution, one of the chemicals is called a solvent, and the other is called a solute. For solutions involving the mixing of chemicals of two different states, the solvent is considered to be the chemical that is of the same state as the resulting solution. For example, when solid sugar is dissolved in liquid water to form a liquid solution, the water is called the solvent, and the sugar is called the solute. If two chemicals are of the same state, the solvent is usually considered to be the component present in the greater amount. In any case, the solvent is the dissolver and the solute is the dissolved component of the solution.

Since solutions are a special type of mixture, the composition of solutions depends on the amount of each component that is used to make the solution. Often there are limits to the amount of solute that can be mixed with a solvent. The amount of solute that will dissolve in a given amount of solvent at a specific temperature is called the solubility of the solute. The solubilities of chemicals in liquid water are sometimes noted as distinguishing properties of the chemical. For example, salt and sugar are soluble in water while vegetable oils and gasoline are not.

In the environment, a chemical is in the solid, liquid, or gaseous state. Chemicals can be changed from one state to another. Such changes of state can be made to occur by changing the environment of the chemical. For example, by heating ice (solid water) it can be made to melt, change from the solid to the liquid state. By heating, liquid water can be boiled or vaporized, changed from the liquid state to the vapor or gaseous state. Under certain conditions some solids can be converted directly from the solid state to the gaseous state. You have probably observed this with a piece of dry ice (solid carbon dioxide). This change is called sublimation.

Cooling or removal of heat can also cause certain changes to occur. A gas can be cooled and caused to change to the liquid state. This is what happens when a window becomes fogged with water droplets that are formed from the water vapor in the air. This change is called condensation. A liquid can be cooled and caused to change to the solid state. This is, of course, what happens when water freezes to form ice. Such a change is called solidification, crystallization, or freezing. The temperature at which certain changes of state occur can be used to characterize and identify chemicals. The temperature at which a solid melts is called the melting point of the solid. For instance, ice melts at about 0 °C. Similarly, the temperature at which a liquid solidifies under specific conditions is called the freezing point of the liquid. For a pure chemical the freezing and melting points are the same. The temperature at which a liquid under specific conditions boils is called the boiling point of the liquid. The observation of freezing, melting, and boiling points is often carried out in the laboratory for purposes of describing chemicals.

Laboratory Procedure

1. APPEARANCES OF CHEMICALS

Several samples of chemicals are on display. Observe the samples and give a description of each including the color. Also give the name of the chemical as given by the label on the sample. Record your observations in the table given on the Report Sheet. (A)

2. MAKING A MIXTURE

Heterogeneous mixtures are made up of distinct parts. These mixtures can often be easily separated. In this exercise you are going to make a mixture and then separate the components by filtration. Filtration involves passing the mixture through a membrane that allows the liquid to pass but retains the solid. Usually, the membranes used in filtration are made of specially prepared paper called filter paper. Such paper has small pores that allow the passage of the liquid only.

To carry out the filtration process you need a funnel supported by a ring attached to a ring stand as shown in Figure 4-1 on the next page. The figure also shows how to fold the filter paper into a cone and how to filter a mixture.

You will need a 50-mL flask and a 100-mL beaker. Set up a funnel as shown in Figure 4-1 and place a folded piece of filter paper in the funnel as instructed in the figure. Add about 25-mL of distilled water to the flask. Obtain a sample of sand, about the size of a marble, and place it on a small piece of paper. Add the sand to the water in the flask. Swirl the flask to mix and describe the appearance of the mixture. (B)

Swirl the mixture in the flask and pour it into the funnel to filter it. Transfer as much of the mixture to the funnel as possible. Allow the water to flow through the filter paper and collect it in the 100-mL beaker. If any sand remains in the flask, hold the flask at an angle over the funnel so that the mouth of the flask is directly over the top of the filter paper. Use a stream of water from your plastic wash bottle to wash any remaining sand into the filter. Allow the water to pass through the filter paper and describe what happened when you filtered the mixture. (C) Remove the filter paper from the funnel and throw it away.

3. MAKING A SOLUTION

For this exercise you will need to set up a funnel for filtration as described in part 2. Fold and place a filter paper cone in the funnel as instructed in Figure 4-1 on the next page. You will need a 50-mL flask and a 100-mL beaker. Place the beaker under the funnel as shown in Figure 4-1. Add about 25 mL of water to the flask. Obtain a sample of potassium permanganate, about the size of a pea, and place it on a clean piece of paper. Pour the sample into the flask of water. Look very closely at the sample as it sinks to the bottom of the flask. Now swirl the flask to mix. Continue swirling until all of the sample dissolves. Describe the appearance of the solution. (D)

Pour the solution into the filter paper and, after all of the liquid passes through the funnel, wash the paper with a stream of water from your plastic wash bottle. Describe what happened when the solution was passed through the filter. (E)

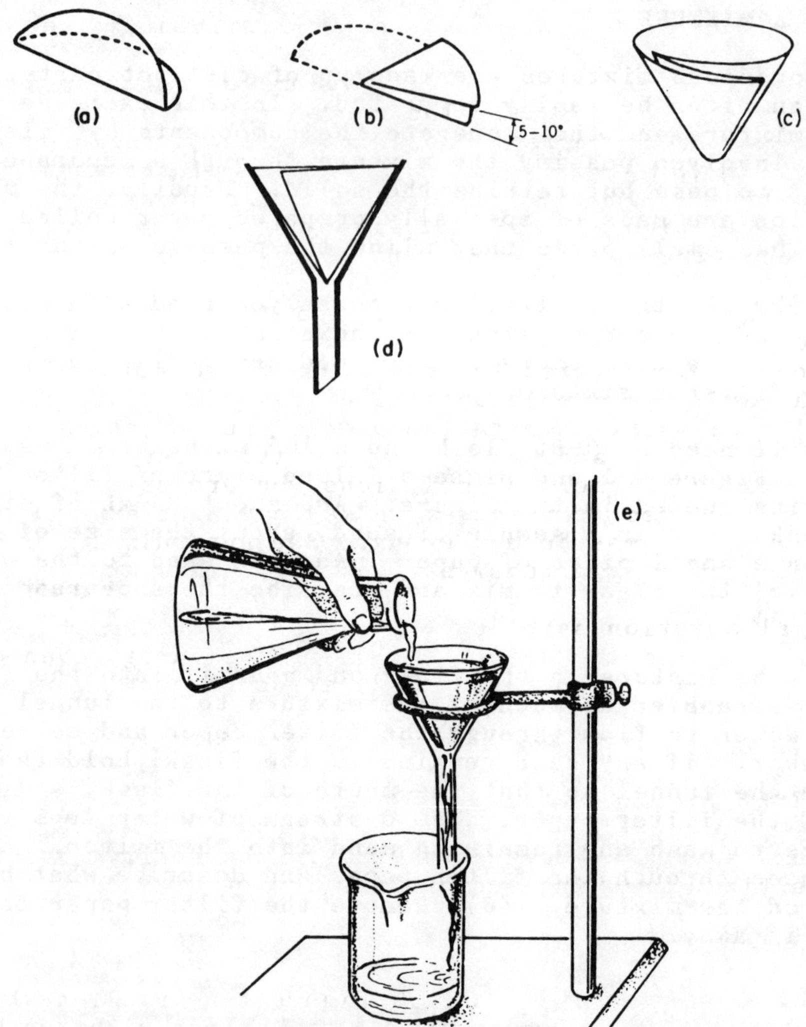

Fig.4-1 The filtration process. (a) Fold in half. (b) Fold again but not exactly in quarters. (c) Open to form a cone. (d) Place cone in funnel. Wet the paper with a small amount of water and seal the paper in the funnel. (e) Support the funnel and carefully pour the mixture to be filtered into the funnel. Never fill the paper cone full of liquid. You may fill it about three-quarters full.

4. SOME SOLUTIONS

When we work with some chemicals in the laboratory, it is convenient to have them dissolved in water as water solutions. Solutions are special kinds of mixtures that are used as easily dispensed sources of certain chemicals. Solutions can be prepared by dissolving one chemical in another. The chemical that dissolves the other chemical is called the solvent. The dissolved chemical is called

the solute. Most solutions that we will use have water as the solvent, but it is possible to have solutions with solvents other than water.

The amount of solute in a unit volume of solution is called the concentration of the solute. The concentration of the solute is usually expressed in terms of molarity which is represented by the capital letter M. Molarity is an expression of the number of moles of solute per liter of solution. You will learn about this concentration term at a later time. In the meantime just look for bottles labeled with the concentration expressed in terms of M.

As you do various experiments in this laboratory manual, you may be instructed to use a solution of specific molarity. For instance, suppose you are instructed to use 10 mL of a 0.1 M sodium chloride, NaCl, solution. You should obtain the desired volume from a bottle of sodium chloride solution with the proper concentration on the label (0.1 M NaCl or 0.1 M sodium chloride). It is very important that you carefully read the labels of solutions you are using. The use of a solution of incorrect concentration may cause an experiment to not work or, in some cases, may be dangerous. See Appendix 2 for a description of the concentrations of common laboratory acids and bases.

Several solution samples are on display in the laboratory. Observe the samples and describe them in the table given on the Report Sheet. (F) Also record the concentration of each solution as given on the label.

5. CHANGES OF STATE

In this exercise you are to measure the boiling point of water and observe melting, freezing and sublimation of some chemicals.

(a) The Boiling Point of Water

Set up the apparatus shown in Figure 4-2 on the next page. Place about 5 mL of distilled water in the tube along with two boiling chips. The thermometer is immersed into the test tube so that the bulb is about 1 cm above the liquid surface but not touching the surface. Do not allow the thermometer to touch the sides of the test tube.

Light a Bunsen burner and, using your hand, raise it to heat the water in the test tube. Carefully heat the water until it begins to boil. Do not allow the liquid to boil out of the tube. Control the heating to obtain a gentle boiling of the water. As the water boils, note that a drop forms on the bulb of the thermometer. When this occurs and the water is gently but continuously boiling, read the temperature registered by the thermometer. This is the boiling point of the water. Allow the water to cool slightly. Then boil it again and obtain another thermometer reading of the boiling point of water. Record your measured boiling point. (G)

(b) Melting, Freezing and Sublimation

Set up an iron ring supported by a ring stand and place a wire gauze on the ring as shown in Figure 1-5. See Experiment 1 for this figure. Fill a 100-mL beaker two-thirds full of water, place it on the wire gauze and heat it with a Bunsen burner.

Obtain a sealed glass tube containing some solid lauric acid and a sealed tube containing some solid iodine. Place both tubes in the beaker of heating water. As the water heats watch the contents of the tubes. Notice the lauric acid tube first. Move it to the side of the beaker and watch the contents. Continue heating until you see the entire solid lauric acid sample melt. Record your observations. (H) After the sample melts, remove the tube from the water. Hold the tube as it cools and watch the contents. Record your observations as the lauric acid freezes. (I)

Shut off the Bunsen burner and remove the tube of iodine from the beaker. Observe the contents and continue observing as the tube cools. Record your observations. (J)

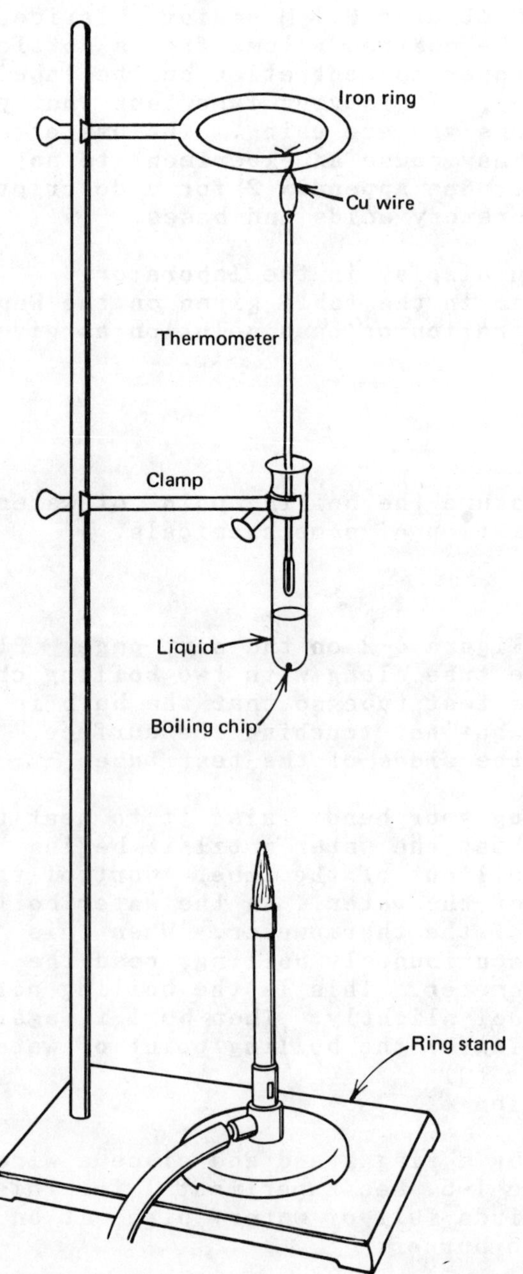

Fig. 4-2 Apparatus used for boiling point determination.

5 Empirical Formula

Objective

The purpose of this experiment is to carry out a chemical reaction that forms a chemical compound and to determine the empirical formula of the compound.

Discussion

The formula of a compound indicates which elements are present in a compound and the relative amounts of these elements. For example, the formula $CaCl_2$ indicates that the compound contains one combined atom of calcium for every two combined atoms of chlorine. From a molar point of view, the formula indicates that a mole of the compound contains two moles of combined chlorine atoms for each mole of combined calcium atoms. This can be expressed as a molar ratio

$$\left(\frac{2 \text{ mole Cl}}{1 \text{ mole Ca}}\right)$$

The molar ratio indicates the subscripts to be used in the formula (i.e., Ca_1Cl_2; the number 1 is omitted when writing the formula).

The simplest formula of a compound can be deduced by determining the amount in grams of each element present in a sample of a compound. This is done by reacting known amounts of elements with one another to form a compound or by separating the elements present in a sample of a compound and determining the amount of each individually. Once the amounts in grams of each element in a compound are known, the number of moles of each element can be calculated. The molar ratios can be found from the number of moles and used to deduce the empirical formula of the compound.

Consider an example of the experimental determination of the empirical formula of a compound. (Do the computations and fill in the blanks in the following discussion.) The example will concern the formation of a compound between gallium, Ga, and oxygen formed by heating a sample of gallium in air. It is found that a 2.79 g sample of gallium forms a product having a mass of 3.75 g. Assuming that the product is the gallium oxygen compound, the mass of oxygen combined with gallium is found by subtracting the gallium mass from the product mass.

Mass of combined oxygen = 3.75 g - 2.79 g = __________

The number of combined moles of each element can be found from the masses.

moles of oxygen: _______________ g $\left(\frac{1 \text{ mole O}}{16.00 \text{ g}}\right)$ = _____________

moles of gallium: $2.79 \text{ g} \left(\frac{1 \text{ mole Ga}}{69.72 \text{ g}} \right) =$ ____________

Dividing the smaller number of moles into the larger will give the molar ratio of the two elements.

$$\left(\frac{\_\_\_\_\_\_\_\_\_\_\_\_ \text{ moles O}}{\text{moles Ga}} \right) = \left(\frac{\_\_\_\_\_\_ \text{ moles O}}{1 \text{ mole Ga}} \right)$$

The formula using this molar ratio is:

Ga_____ O_____

Since the ratio is not a whole number, each subscript should be multiplied by the number __________. This gives the final empirical formula of:

Ga_____ O_____

The empirical formula of a magnesium-oxygen compound is to be determined in this experiment. The magnesium-oxygen compound is made by reacting magnesium metal with oxygen in the air. This will be accomplished by weighing a sample of magnesium and reacting it with oxygen in a closed crucible. The mass of the product is then determined, and the mass of oxygen that reacted with the magnesium can be found by subtracting the magnesium mass from the product mass. One problem is that at high temperatures some nitrogen will also react with the magnesium. This problem is overcome by reacting the magnesium-nitrogen compound with water and heating to convert to the magnesium-oxygen compound.

Laboratory Procedure

Remember to always wear safety goggles when heating chemicals. Wash, rinse and dry your crucible and cover. Place the crucible with the cover on a clay triangle supported on a ring stand as shown in Figure 5-1 on the next page. Place your Bunsen burner and the ring so that the hot part of the flame reaches the bottom of the crucible. After heating, remove the burner and allow the crucible to cool to room temperature. While the crucible is cooling, obtain about 50 cm of magnesium ribbon. (If necessary, clean any scale off the ribbon with steel wool or sand paper and wipe the ribbon clean with a paper towel or tissue.) Fold the magnesium sample into an accordion-like shape with about 2 cm length folds. It should be folded so that it will fit into the crucible.

Weigh the cooled crucible and cover to the nearest 0.01 g and record the mass on the report sheet. (B) Place the folded magnesium sample into the bottom of the crucible and weigh the crucible, magnesium and cover. Record the mass. (A) Cover the crucible and place it back on the clay triangle. Heat the crucible gently at first by moving the flame back and forth under the crucible for a few minutes. Then heat the crucible with the full flame for 15 minutes. (DANGER: Do not heat the crucible with the cover off and do not remove the cover while heating.) After 15 minutes of heating, allow the crucible to cool for about 5 minutes and then, using your crucible tongs, adjust the cover so that there is a very small opening to the

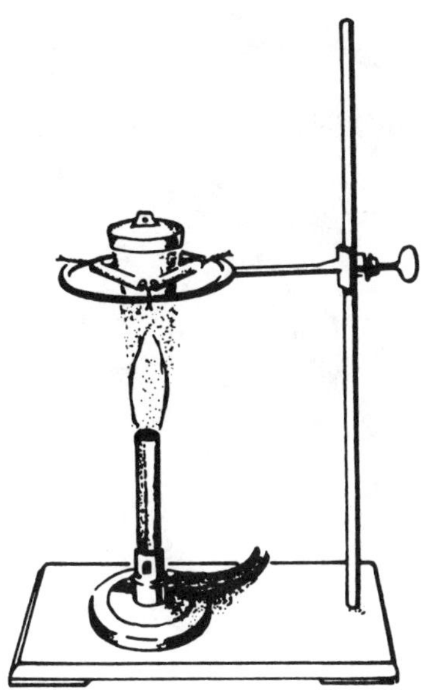

Fig. 5-1

crucible. (See Fig. 5-2.) Heat the crucible with a strong flame for 10 more minutes, remove the flame, and cool for a few minutes. Cautiously raise the cover with your crucible tongs and heat the crucible strongly. If the contents of the crucible glow brightly and white smoke rises, place the cover on and heat for 5 minutes. Repeat the process of raising the cover, observing the product and replacing the cover until all of the magnesium has appeared to react. This will be apparent when the contents do not glow brightly when red hot. After you think the magnesium has reacted, heat strongly for a few minutes while holding the cover off. Replace the cover and allow to cool to roor temperature.

Fig. 5-2

After you are sure the crucible is cool, add 10 to 15 drops of distilled water to the crucible, cover and gently heat for 5 minutes. Water is added to convert any magnesium-nitrogen compounds to magnesium-oxygen compounds. Allow the crucible to cool to room temperature. The crucible will be cool enough to weigh if you can touch it and it does not feel warm. Finally, weigh the crucible, cover and product to the nearest 0.01 g and record the mass. (D)

Use the data recorded on the report sheet to calculate the mass of magnesium and the mass of oxygen. Use these masses to determine the empirical formula of the magnesium-oxygen compound.

Fig. 3-1

crucible. (See Fig. 3-2.) Heat the crucible with a strong flame for 10 more minutes, remove the flame and cool for a few minutes. Cautiously raise the cover with tongs [illegible] and heat the [illegible] strongly. [illegible] the contents of the crucible glow brightly and white smoke rises, replace the cover and heat for 10 minutes. Repeat [illegible] raising the cover, observing the product and replacing the cover [illegible] the [illegible] correct. [illegible] be [illegible] when the contents do not glow brightly when [illegible] you [illegible] the magnesium has reacted. Heat strongly for a few minutes while holding the cover off. Replace the cover and allow to cool to room temperature.

After you are sure the crucible is cool, add 10 to 15 drops of distilled water to the crucible. Cover and gently heat for [illegible] minutes. Water was added to convert any magnesium [illegible] compounds to magnesium oxide [illegible]. Then allow the crucible to cool to room temperature. The crucible will be cool enough to weigh if you can touch it and it does not feel warm. Finally, weigh the crucible, cover and product to the nearest 0.01 g. [illegible]

Use the data recorded on the report sheet to calculate the mass of magnesium and the mass of oxygen. Use these masses to determine the empirical formula of the magnesium-oxygen compound.

6 Formula Worksheet

6

Name ________________

Lab Section ________________

Due Date ________________

The formula of a compound contains information about the compound that can be used in several types of chemical computations. The following exercises allow you to practice the use of formulas.

1. The formula of sodium acetate is $NaC_2H_3O_2$. List the names of the elements in this compound and indicate the number of combined atoms of each element in the compound.

2. Find the molar mass of sodium acetate to four digits and express the molar mass as a factor.

3. Use the molar mass of sodium acetate as a factor to:

   a. determine the number of moles of $NaC_2H_3O_2$ in a 12.9 g sample.

   b. determine the mass of 1.6×10^{-2} moles of $NaC_2H_3O_2$.

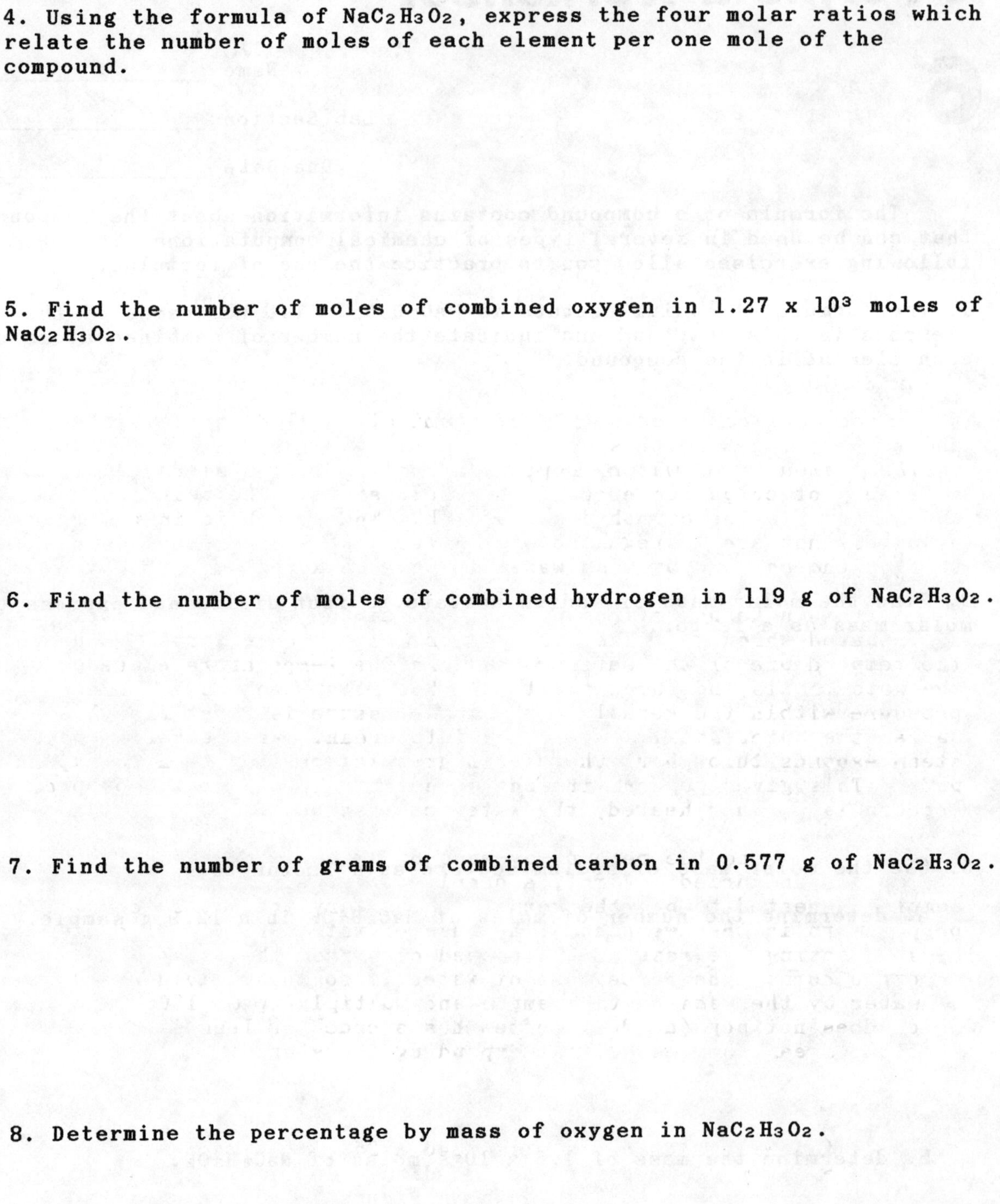

4. Using the formula of $NaC_2H_3O_2$, express the four molar ratios which relate the number of moles of each element per one mole of the compound.

5. Find the number of moles of combined oxygen in 1.27×10^3 moles of $NaC_2H_3O_2$.

6. Find the number of moles of combined hydrogen in 119 g of $NaC_2H_3O_2$.

7. Find the number of grams of combined carbon in 0.577 g of $NaC_2H_3O_2$.

8. Determine the percentage by mass of oxygen in $NaC_2H_3O_2$.

7 Popcorn: Water in a Mixture

Objective

The purpose of this experiment is to determine the percentage by mass water in popcorn.

Discussion

Foods are composed mainly of chemicals called carbohydrates, fats and proteins along with small amounts of vitamins, minerals and variable amounts of water. Popcorn kernels are corn seeds consisting of a variety of compounds encased in a hard shell. The main component of the mixture is the carbohydrate starch. The starch is in the form of granules that are impregnated with water. This experiment is designed to find the percent by mass water in popcorn kernels.

Within a kernel of popcorn a small amount of water is intimately distributed throughout the starch granules. When a kernel is heated, the temperature of the water rises. As the temperature exceeds 100 °C, the water boils and turns to steam. The steam expands and exerts a pressure within the kernel. When the pressure is great enough, it causes the outer shell of the kernel to break. As the kernel pops, the steam expands throughout the starch granules causing them to expand or puff. This gives popcorn its characteristic form. As the popped popcorn is further heated, the water can escape.

To determine the percent-by-mass of water in popcorn, the kernels are popped and dried. First, a sample of kernels is weighed. The sample is heated to pop the kernels and to dry the popped corn. The popped corn is then weighed. The mass of water driven off can be found by subtracting the mass of the popped corn from the mass of the unpopped corn. The percentage of water is found by dividing the mass of water by the mass of the sample and multiplying by 100. A kernel which does not pop (a "dud") often has a crack or leak in the shell so that the steam escapes without expanding the starch.

Laboratory Procedure

For this experiment you will need a 250-mL beaker, an evaporating dish and a watch glass cover for the dish. Set up an iron ring supported by a ring stand as shown in Figure 7-1 on the next page. Place a wire gauze on the ring.

Obtain 15 kernels of unpopped corn on a clean piece of paper or in a small beaker. Weigh the evaporating dish and watch glass cover to the nearest 0.01 g. (B) Place the kernels in the dish, cover it and weigh to the nearest 0.01 g. (A)

Position a Bunsen burner under the ring supported by the ring stand and adjust the level of the ring so that the flame will reach the gauze. (See Figure 7-1.) Do not light the burner. Place the 250-mL beaker on the gauze and place the covered evaporating dish on top of the beaker as shown in the figure. The beaker will serve as an air bath to heat the evaporating dish. This avoids direct contact of the flame and the dish.

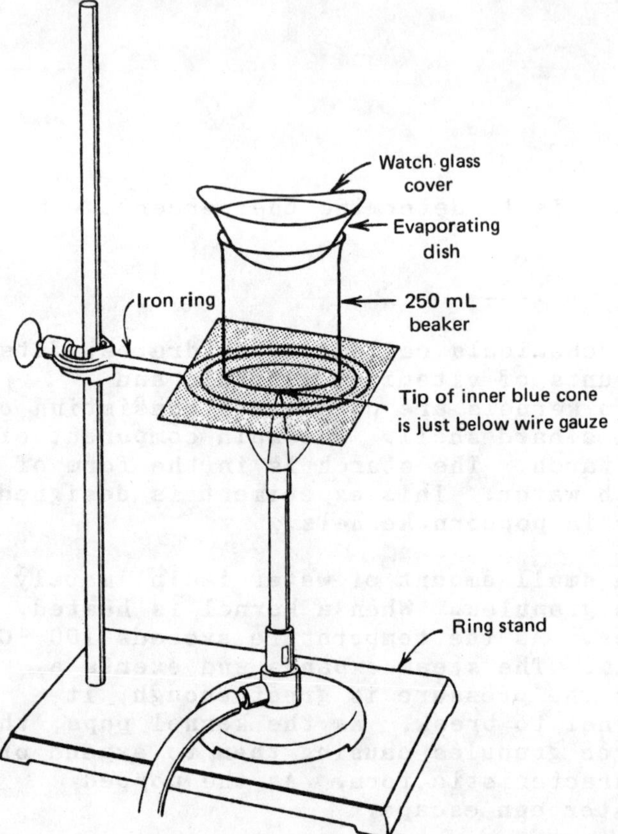

Fig. 7-1 Apparatus for Air Bath.

Light the burner and make sure that the hot part of the flame is in contact with the bottom of the beaker. Allow the system to heat until the kernels pop. As soon as they all pop, shut off the heat. Too much heating may cause the popcorn to scorch or burn. If this happens, let the apparatus cool and start over with a new sample of kernels. If some of the kernels are "duds", just continue with the experiment.

Allow the dish to cool to room temperature. When it is cool enough to touch, you can move the dish to the top of the desk so that it will cool faster. After the covered dish has reached room temperature, weigh it to 0.01 g. (C) Clean the dish and repeat the experiment to obtain a second set of data.

The mass of water driven off from the popped corn can be found by subtracting the mass of the dish with the popped corn (C) from the mass of the dish with the unpopped corn (A). Determine the mass of water. (D) Find the mass of the unpopped kernels by subtracting the mass of the covered dish (B) from the mass of the dish and the unpopped corn. (A). (E) Use the data to calculate the percentage by mass of water in the popcorn by dividing the mass of water by the mass of the unpopped kernels and multiplying by 100. (F) After calculating the percentage for each sample, calculate the average of your two results. (G) If your two results seem too far apart, consult your instructor.

7

REPORT SHEET

Name ________________

Lab Section ________________

Due Date ________________

EXPERIMENT 7

Experimental Data

(A Mass dish, cover and kernels	__________	__________	__________
(B) Mass dish and cover	__________	__________	__________
(C) Mass dish, cover and popped corn	__________	__________	__________
(D) Mass of water (A) - (C)	__________	__________	__________
(E) Mass of sample (A) - (B)	__________	__________	__________

(F) Percentage water (Show setup and answer.)

(G) Average percent-by-mass water

7

QUESTIONS

Name ______________________________

Lab Section ______________________

Due Date _________________________

EXPERIMENT 7

1. Why does unpopped popcorn pop?

2. Using the data for the grams of water obtained in your experiment and the fact that there were 15 kernels of corn, calculate the number of grams of water per one kernel of popcorn.

3. Your answer for the percent by mass of water in popcorn can be expressed as the number of grams of water per 100 grams of popcorn. Using your percentage in this form determine the number of grams of water that is contained in 1.0 pounds of unpopped corn.

4. Using your answer for question 1 which expresses the number of grams of water per popcorn kernel as a conversion factor and your answer for question 2, determine the approximate number of kernels that are in 1.0 pounds of unpopped corn.

8 Hydrates: Water in a Complex Compound

Objective

The purposes of this experiment are to study the behavior of some hydrates and to determine the percentage of water by mass in a hydrate.

Discussion

Water is one of the most common chemicals found on earth. It is found underground, in lakes, in rivers, and in the oceans. It is even found in the atmosphere as clouds and water vapor. Water is very useful and is fundamental to life processes. Water is sometimes found as a component of compounds called hydrates. Hydrates can be formed by the inclusion of water into the crystals of a solid. Hydrates are interesting kinds of complex compounds that include water combined with another compound. Some minerals are hydrates. Some examples are gypsum, borax and Epsom salts. Hydrates are complex compounds because they involve water combined with another compound. That is, they are compounds composed of two other compounds one of which is water. In fact, the word hydrate means "combined with water." A hydrate is a complex compound containing a specific number of moles of water per mole of the compound. Formulas for hydrates are written by stating the formula of the compound other than water followed by a dot and the formula of water preceded by a number showing the number of moles of water per mole of compound. For example, a hydrate of calcium chloride contains two moles of water per mole of compound. Thus, the formula is $CaCl_2 \cdot 2H_2O$. A few examples of hydrates are:

$CuSO_4 \cdot 5H_2O$	Copper(II) sulfate pentahydrate
$CaCl_2 \cdot 2H_2O$	Calcium chloride dihydrate
$MgSO_4 \cdot 7H_2O$	Magnesium sulfate heptahydrate
$Na_2CO_3 \cdot 10H_2O$	Sodium carbonate decahydrate

The water present in hydrates is called water of hydration. It is sometimes possible to remove the water of hydration or dehydrate the hydrate by gently heating a sample. Heating a hydrate produces water and the anhydrous (without water) solid compound. For example

$$CaCl_2 \cdot 2H_2O \xrightarrow{\text{heat}} \underset{\text{anhydrous calcium chloride}}{CaCl_2} + H_2O$$

The percentage of water by mass in some hydrates can be determined by dehydrating a known amount of the hydrate and then determining the mass of the resulting anhydrous solid. The mass of water is found by subtracting the mass of anhydrous solid from the mass of the original sample of hydrate:

$$\text{Mass } H_2O = \text{mass hydrate} - \text{mass anhydrous solid}$$

The percentage of water can be found by dividing the mass of water by the mass of the original sample of hydrate and multiplying by 100.

$$\text{Percent } H_2O = \left(\frac{\text{mass } H_2O}{\text{mass hydrate}}\right) \times 100$$

Some anhydrous solids and other chemicals can combine with water in the atmosphere to form hydrates. When a chemical absorbs water from the atmosphere, it is said to be hydroscopic. Sometimes hydroscopic chemicals are used as drying agents or dessicants. Some chemicals are so hydroscopic that they absorb enough water from the atmosphere to form a solution. Such chemicals are said to be deliquescent. In fact, deliquescent chemicals have to be stored in tightly sealed containers or they will dissolve in the water they absorb from the atmosphere. A few hydrates actually spontaneously lose water of hydration when open to the atmosphere. Such a process is called efflorescence and such hydrates are called efflorescent chemicals.

Laboratory Procedure

1. DELIQUESCENCE AND EFFLORESCENCE

Place a few crystals of sodium sulfate decahydrate, $Na_2SO_4 \bullet 10H_2O$, on a sheet of paper or a watch glass. Place a few crystals of calcium chloride, $CaCl_2$, on the same sheet but as far from the other crystals as possible. Allow the crystals to remain on the sheet for about one hour and then record any changes you observe. (Go on with the rest of the experiment and return to this part later.) Which of the chemicals is deliquescent? Which of the chemicals is efflorescent? (A)

2. EFFECT OF HEATING A HYDRATE

Place a few crystals of copper(II) sulfate pentahydrate in a dry test tube. Hold the test tube almost horizontally and gently heat the crystals by passing the tube through a Bunsen flame. Note any changes. When no further changes occur, allow the tube to cool. After cooling, add a few drops of water. Note any changes. Feel the tube and note any temperature change. Describe your observations. (B)

3. THE PERCENT OF WATER BY MASS IN A HYDRATE

Clean a crucible with a cover and place the covered crucible in a clay triangle supported on a ring stand. (See Fig. 8-1.) Carefully heat the crucible and cover for a few minutes to make sure they are dry. Allow to cool to room temperature and determine the mass of the covered crucible to the nearest 0.01 g. (D) Obtain a sample of hydrate. Place about 3 g of the hydrate in the crucible, cover, and weigh to the nearest 0.01 g. (C)

Put the crucible back on the clay triangle and drive off the water by gently heating the crucible for about 15 minutes. Use a medium flame to heat the crucible and do not allow the crucible to become red hot. This can be done by making sure that the tip of the flame does not touch the bottom of the crucible as shown in Fig. 8-1. After heating, cool to room temperature and determine the mass to the nearest 0.01 g. (E) Gently heat the crucible for another 5 minutes, cool and weigh. (E) If the first and second weighings do not agree within 0.03 g, reheat for 5 minutes and weigh again. (E) Repeat until subsequent weighings agree.

The mass of water which was contained in the sample can be found by subtracting the final mass of the heated crucible from the mass of the crucible with the hydrate. (F) The mass of the hydrate sample can be found by subtracting the mass of the empty crucible from the mass of the crucible with hydrate. (G) The percentage of water by mass in the hydrate can be found by dividing the mass of water by the mass of the hydrate sample and multiplying by 100. (H)

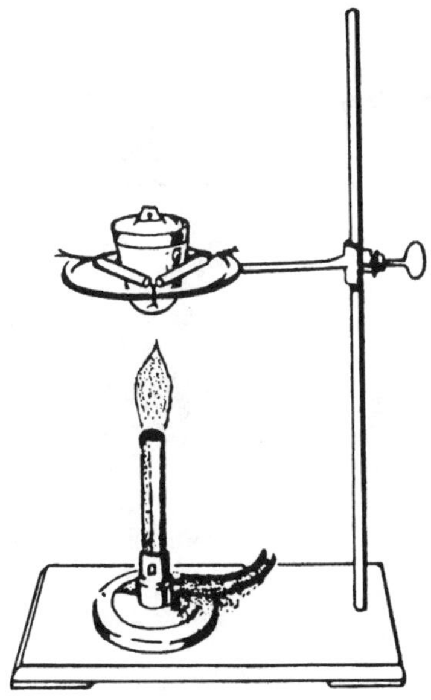

Fig. 8-1

4. THE FORMULA OF THE HYDRATE (OPTIONAL)

Show your percentage water results to your instructor and he or she will give you the formula of the parent compound contained in the hydrate. From the formula you can find the number of grams per mole of the parent compound. To determine the formula of the hydrate, calculate the percent of parent compound in the hydrate by subtracting the percent water from 100.

______________(I)

Based on this percentage, how many grams of parent compound would be contained in 100 g of hydrate?

______________(J)

Use the molar mass of the parent compound to find the number of moles of this compound.

______________(K)

Using the percentage water, how many grams of water would be contained in 100 g of the hydrate?

______________(L)

Use the molar mass of water to find the number of moles of water.

______________(M)

The formula of the hydrate can be found by determining the number of moles of water per mole of the parent compound. This is done by dividing the number of moles of water by the number of moles of parent compound.

______________(N)

The formula of the hydrate is

__________ • _______H_2O (O)

8

REPORT SHEET

Name _______________

Lab Section _______________

Due Date _______________

EXPERIMENT 8

1. Deliquescence and Efflorescence

(A)_______________efflorescent_______________deliquescent

2. Effect of Heating a Hydrate

(B)

3. The Percent Mass Water in a Hydrate

Mass crucible and hydrate _____________(C)

Mass crucible _____________(D)

Mass heated crucible

____________ ____________ ____________ _____________(E)

Mass water (C) minus (E) _____________(F)

Mass hydrate (C) minus (D) _____________(G)

Percent by mass water ____________ 100 =_____________(H)

4. Formula of Hydrate

Percent water _____________

Percent parent compound _____________(I)

Grams parent compound _____________(J)

Molar mass parent compound _____________

Moles parent compound

____________ ____________ = _____________(K)

Grams water _____________(L)

Moles water

_____________ ____________ = _____________(M)

Moles of water per mole of parent

____________ = _____________(N)

Formula of hydrate ____________ • __________H_2O(O)

8

QUESTIONS

Name ________________________

Lab Section _________________

Due Date ____________________

EXPERIMENT 8

1. Explain how each of the following factors would affect the determination of the percentage of water in a hydrate. Indicate whether the factor will give a high or low percentage and explain why.

(a) The hydrate was not heated sufficiently and some of the original hydrate is present at the end of the experiment.

(b) The dehydrated hydrate absorbs some water from the atmosphere before it is weighed at the end of the experiment.

2. Plaster of Paris used in plaster casting is made by gently heating gypsum, $CaSO_4 \cdot 2H_2O$. Actually Plaster of Paris is a also a hydrate but one that contains less water of hydration than gypsum. When water is added to powdered Plaster of Paris, it reforms solid gypsum. If Plaster of Paris is a hydrate of calcium sulfate containing 6.18% water, what is the formula of the Plaster of Paris hydrate?

3. Using your formula for Plaster of Paris obtained in question 2 and the formula for gypsum shown in question 2, write a balanced equation that shows the reaction between Plaster of Paris and water to form gypsum.

4. A desiccant is a chemical that can be used to keep a closed container or a confined space free from excess water vapor. Typical desiccants are silica gel and calcium chloride. Explain how desiccants work in terms of hydrate formation.

9 Analysis of Vinegar

Objective

The purpose of this experiment is to carry out a chemical analysis of the acetic acid content of vinegar.

Discussion

Vinegar has been used for centuries as a sour food flavoring agent. Literally, vinegar means "sour wine" which suggests how it was first discovered. Vinegar is a water solution of acetic acid, $HC_2H_3O_2$. In this experiment you are going to analyze a sample of vinegar to find the percent by mass of acetic acid in the vinegar. To find the amount of acetic acid in a sample of vinegar, a solution of sodium hydroxide, NaOH, is added and the following reaction occurs:

$$HC_2H_3O_2(aq) + NaOH(aq) \longrightarrow NaC_2H_3O_2(aq) + H_2O$$

The solution of NaOH contains a known number of moles of dissolved NaOH per gram of solution. If we determine the number of grams of NaOH solution needed to react with the acetic acid in the vinegar, we can determine the number of moles of $HC_2H_3O_2$. This can be done by using the fact that 1 mole of NaOH is needed to react with 1 mole of $HC_2H_3O_2$. Once we know the number of moles of acetic acid, we can find the mass of the acid using its molar mass. To illustrate, suppose we analyze a vinegar solution using a NaOH solution which contains (0.00128 moles NaOH/g of solution). It is found that it requires 12.32 g of a NaOH solution to react with the acetic acid in a vinegar sample. The number of moles of NaOH involved in the reaction is found by multiplying the mass of the solution used by the number of moles of NaOH per gram of solution:

$$12.32\text{ g}\left(\frac{0.00128\text{ mole NaOH}}{1\text{ g}}\right)$$

Next, the moles of acetic acid in the vinegar can be calculated by using the fact that 1 mole of $HC_2H_3O_2$ reacts with 1 mole of NaOH.

$$12.32\text{ g}\left(\frac{0.00128\text{ mole NaOH}}{1\text{ g}}\right)\left(\frac{1\text{ mole }HC_2H_3O_2}{1\text{ mole NaOH}}\right)$$

Once we know the number of moles of acetic acid we can calculate the mass of the acid by multiplying by the molar mass:

$$12.32\text{ g}\left(\frac{0.00128\text{ mole NaOH}}{1\text{ g}}\right)\left(\frac{1\text{ mole }HC_2H_3O_2}{1\text{ mole NaOH}}\right)\left(\frac{60.05\text{ g}}{1\text{ mole }HC_2H_3O_2}\right)$$

$$= 0.947\text{ g}$$

Laboratory Procedure

In the experiment, NaOH and vinegar solution will be mixed. To determine when sufficient NaOH has been added, we place an indicator, named bromthymol blue, in the solution. Bromthymol blue turns blue in the presence of NaOH and is yellow in vinegar. If bromthymol blue is added to the solution mixture and the NaOH solution is slowly added until the blue color occurs, this will indicate that the proper amount of NaOH has been added.

For this experiment you will need a 100-mL beaker, a 50-mL flask, a 50-mL beaker and a stirring rod. Obtain about 20 mL of the NaOH solution provided in a clean, dry 100-mL beaker. Record the composition of this solution. (A) CAUTION: Sodium hydroxide solutions are dangerous if they contact your skin or eyes. Handle NaOH solutions with care. Weigh a clean, dry 50-mL beaker to 0.01 g. (C) Pour the 20 mL sample of NaOH solution into the beaker and weigh the beaker and solution to the nearest 0.01 g. (B) Place about 20 mL of vinegar solution in a clean, dry 50-mL flask. Weigh the flash with the vinegar to the nearest 0.01 g. (E)

Add 1 or 2 drops of bromthymol blue indicator to the beaker of NaOH solution. Carefully pour the vinegar solution into the beaker of NaOH solution until the solution becomes yellow. Use a stirring rod to aid you in pouring the vinegar solution as shown in Figure 9-1. CAUTION: Do not spill any of the vinegar or let it dribble down the sides of the flask. After adding the vinegar, weigh the flask to the nearest 0.01 g. (F) Next, weigh a dropper bottle containing the NaOH solution to the nearest 0.01 g. (H)

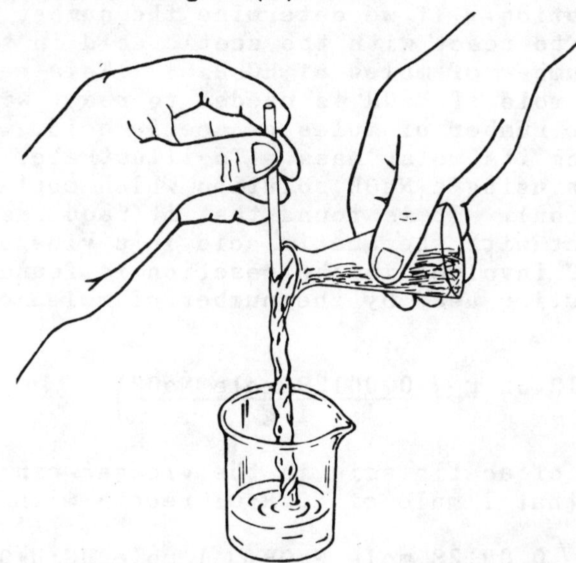

Fig. 9-1 Hold a glass rod flat against the mouth of the flask. Carefully pour the liquid down the rod.

Carefully add the NaOH solution to the yellow NaOH-vinegar mixture, drop by drop and mix the solution after each drop, until the mixture turns a definite blue color. As you add the drops you can mix with the stirring rod. Do not spill any drops and do not add too many drops. After adding the proper amount of NaOH solution, weigh the dropper bottle to 0.01 g. (I) Rinse and dry your glassware and repeat the experiment using another sample.

Using the data in the data table on the Report Sheet, calculate the mass of the vinegar sample (G), the mass of the NaOH solution in the beaker (D), and the mass of NaOH solution from the dropper bottle. (J) Add the two NaOH solution masses to obtain the total mass of NaOH solution used (K). Use your data to calculate the mass of acetic acid in your vinegar sample (L). Calculate the percent by mass of acetic acid in the vinegar by dividing the mass of acetic acid (L) by the mass of the vinegar sample (G) and multiplying by 100 (M). If the results of your percentages agree within 0.1, determine the average of the percentages. If your percentages are not close enough, consult your instructor.

10. Using the data in the data table on the Report Sheet, compute the mass of the vinegar sample (G), the mass of the NaOH solution in the beaker (H), and the mass of NaOH solution from the dropper bottle (I). Add the two NaOH solution masses to obtain the total mass of NaOH solution used (J). Use your data to calculate the mass of acetic acid in your vinegar sample (K). Calculate the percent by mass of acetic acid in the vinegar by dividing the mass of acetic acid (K) by the mass of the vinegar sample (G) and multiplying by 100 (L). If the results of your percentages agree within 0.1, determine the average of the percentages. If your percentages are not close enough, consult your instructor.

9

REPORT SHEET

Name ________________

Lab Section ________________

Due Date ________________

EXPERIMENT 9

Data Table

(A) Composition of NaOH solution as provided ____________________

(B) Mass of beaker + NaOH ____________________

(C) Mass of 50-mL beaker ____________________

(D) Mass of NaOH in beaker ____________________

(E) Mass of flask + vinegar ____________________

(F) Mass of flask - vinegar ____________________

(G) Mass of vinegar ____________________

(H) Mass of dropper bottle ____________________

(I) Mass dropper bottle - NaOH ____________________

(J) Mass of NaOH from dropper ____________________

(K) Total mass NaOH (D) + (J) ____________________

Calculations: Show setup and answers.

(L) Mass of acetic acid in samples

(M) Percent by mass acetic acid in vinegar

9

QUESTIONS

Name ______________________________

Lab Section ______________________

Due Date _________________________

EXPERIMENT 9

1. Two brands of vinegar are analyzed by mixing samples with a sodium hydroxide solution containing (0.00132 moles NaOH/1 g solution). Using the following data, determine which of the brands has the higher percent by mass acetic acid.

	mass of vinegar sample	mass of NaOH solution needed
Brand X	30.25 g	16.04 g
Fred's Deluxe Vinegar	22.53 g	12.38 g

2. When a vinegar solution is analyzed by mixing it with a sodium hydroxide solution of known composition, tell how each of the following factors will affect the calculated number of grams of acetic acid. That is, would the calculated number of grams be greater than it should be, less than it should be, or not affected. Give an explanation of your answer.

(a) Some vinegar solution is spilled while it is being added to the sodium hydroxide solution.

(b) Too much sodium hydroxide solution is added to the vinegar solution.

10 The Detection of Common Ions

Objective

The purposes of this experiment are to perform chemical tests used for the identification of some common ions and to use the tests to detect ions in some familiar chemicals.

Discussion

Chemical spot tests or qualitative tests are chemical reactions used to detect the presence of a specific element or ion. Spot tests are a form of qualitative analysis that reveal the presence of a chemical species in a mixture. This experiment includes some qualitative analysis tests for common ions.

Typically, a qualitative analysis test involves a chemical reaction that is unique for a specific ion. Selected chemicals are added to a solution to be tested for a given ion. If the ion is present, a chemical reaction occurs that gives a noticeable product. The unique reaction usually produces a solid product of notable color or form, or a gas that can be identified.

A qualitative analysis test is run by adding specific amounts of test chemicals to the solution or substance to be tested. If the unique reaction occurs, this indicates the presence of the ion. If no unique reaction occurs, the absence of the ion is indicated.

The common ions for which spot tests are given in this experiment are described below:

1. Ammonium ion, NH_4^+, is a common, positively charged polyatomic ions. Do not confuse the ammonium ion with the neutral compound ammonia, NH_3.

2. Chloride ion, Cl^-, is the very common simple ion formed by chlorine.

3. Carbonate ion, CO_3^{2-}, and hydrogen carbonate ion, HCO_3^-, are chemically related polyatomic ions including carbon.

4. Sulfate ion, SO_4^{2-}, a common sulfur-containing polyatomic ion.

5. Nitrate ion, NO_3^-, the most common nitrogen-containing polyatomic ion.

6. Phosphate ion, PO_4^{3-}, a common polyatomic ion of phosphorus.

Laboratory Procedure

1. QUALITATIVE ANALYSIS TESTS FOR COMMON IONS

As you carry out the following tests, be aware of any changes that occur when the chemicals are mixed. Be sure to use the amounts and volumes of chemicals and solutions as directed. Record your observations and indicate the results in each test in the space provided and on the report sheet. Each test is first run on a sample that is known to contain the ion. This allows you to see what a positive test looks like. Then the test is run on a sample of a substance that may contain the ion. Use small test tubes unless otherwise instructed.

A. AMMONIA, NH_3, AND AMMONIUM ION, NH_4^+ First, a solution will be tested for the presence of ammonia. Pour about 3 mL of household ammonia solution into a 50 mL beaker. CAUTION: Avoid breathing the fumes from the ammonia solution. Moisten a piece of red litmus paper with distilled water and hold it over the beaker so that it is close to, but not touching, the solution. Record your observations. (A) The NH_3 gas contacting the litmus paper causes the color change.

Compounds containing the ammonium ion will not directly give a positive ammonia test. However, compounds or solutions containing ammonium ion can be treated with a sodium hydroxide solution to convert the ammonium ion to ammonia.

$$NH_4^+ + OH^- \longrightarrow NH_3 + H_2O$$

The ammonia formed can be detected by holding red litmus above the solution. The ammonia will turn the red litmus blue.

Test for ammonium ion, NH_4^+:

1. Place a small sample or 1 to 2 mL of solution to be tested in a test tube.

2. Add about 2 mL of a 6 M sodium hydroxide solution.

3. Hold moistened red litmus paper in the mouth of the test tube but do not allow the paper to touch the sides of the tube. It may be necessary to warm the tube, but do not let the solution boil. Do not breathe the fumes.

Using the ammonium ion test:

a. Place about 1 mL of a 1 M ammonium chloride, NH_4Cl, solution in a small test tube and test for ammonium ion. Describe your results. (B) You should obtain a positive test for NH_4^+.

b. Place a sample of commercial fertilizer about the size of a pencil eraser in a small test tube. Test the sample for ammonium ion. You may have to gently heat the tube without boiling. Results. (C)

B. CHLORIDE ION, Cl^- Chloride ion present in a soluble substance can be detected by dissolving the substance, adding nitric acid and then adding some silver nitrate, $AgNO_3$, solution. If chloride ion is present, a characteristic white solid is formed. This solid is silver chloride, AgCl, formed by the reaction:

$$Ag^+ + Cl^- \longrightarrow AgCl$$

Test for chloride ion, Cl^-:

1. Place a small sample or 1 to 2 mL of solution to be tested in a test tube.

2. Add 2 or 3 drops of 6 M nitric acid.

3. Add 5 or 6 drops of 0.1 M silver nitrate solution.

4. Note whether or not a white solid forms.

Using the chloride ion test:

a. Place a sample of table salt about the size of a pencil eraser in a test tube. Add 1 to 2 mL of distilled water and test for chloride ion. Record your observations. (D) You should get a positive test for Cl^-.

Test a sample of tap water for chloride. Result. (E)

C. CARBONATE ION, $CO_3{}^{2-}$, OR HYDROGEN CARBONATE ION, $HCO_3{}^-$ Substances that contain carbonate ion or hydrogen carbonate ion react with acid solutions to form carbon dioxide gas. The carbon dioxide gas usually bubbles out of solution. The reactions producing carbon dioxide are:

$$HCO_3{}^- + H_3O^+ \longrightarrow 2H_2O + CO_2$$

$$CO_3{}^{2-} + 2H_3O^+ \longrightarrow 3H_2O + CO_2$$

The presence of carbon dioxide can be confirmed as follows. A length of copper wire is bent to form a small loop on the end. The wire is dipped into a solution of barium hydroxide, $Ba(OH)_2$, to obtain a drop in the loop. The wire in then carefully inserted into the tube in which the gas is produced. If the drop turns cloudy as a white solid is formed, the gas is carbon dioxide. The while solid is barium carbonate formed by the reaction:

$$Ba^{2+} + 2OH^- + CO_2 \longrightarrow BaCO_3 + H_2O$$

Test for carbonate ion, CO_3^{2-}, or hydrogen carbonate ion, HCO_3^-:

1. Place a small sample or 1 or 2 mL of solution to be tested in a test tube.

2. Add 2 to 3 mL of 6 M sulfuric acid. (Be careful with sulfuric acid!)

3. Watch for the evidence of gas bubbles and test any gas being evolved by inserting an wire loop holding a drop of a barium hydroxide solution.

4. Note whether or not the barium hydroxide solution turns cloudy.

Using the carbonate ion test:

a. Place a pencil eraser-sized sample of baking soda ($NaHCO_3$, sodium hydrogen carbonate) in a test tube. Test for hydrogen carbonate ion. Describe your results. (F) You should get a positive test for HCO_3^-.

b. Place a few pieces of marble in a test tube and test for carbonate ion. Describe your results. (G)

D. SULFATE ION, SO_4^{2-} Sulfate ion present in a soluble substance can be detected by dissolving the substance, adding nitric acid and then adding some barium chloride, $BaCl_2$, solution. If sulfate ion is present, a characteristic white solid is formed. This solid is barium sulfate, $BaSO_4$, formed by the reaction:

$$Ba^{2+} + SO_4^{2-} \longrightarrow BaSO_4$$

Test for sulfate ion, SO_4^{2-}:

1. Place a small sample or 1 or 2 mL of solution to be tested in a test tube.

2. Add 2 or 3 drops of 6 M nitric acid. Test with blue litmus to see if the solution turns litmus red. If not add more acid drops.

3. Add 5 or 6 drops of 0.2 M barium chloride, $BaCl_2$, solution.

4. Note whether or not a white solid forms.

Using the sulfate ion test:

a. Place about 1 mL of 0.01 M sodium sulfate, Na_2SO_4, in a test tube and test for sulfate ion. Record your observations. (H) You should get a positive test for SO_4^{2-}.

b. Test a pencil eraser-sized sample of Epsom salt for sulfate ion. Record your results. (I)

E. NITRATE ION, NO_3^- It is possible to detect nitrate ion in soluble substances or solutions by adding sodium hydroxide solution and aluminum metal. A reaction occurs that changes the nitrate ion to ammonia. The reaction is:

$$3NO_3^- + 8Al + 5OH^- + 18H_2O \longrightarrow 8Al(OH)_4^- + 3NH_3$$

The ammonia can be detected using red litmus as was done in the ammonium ion test. Of course, ammonia and ammonium ion in a solution will interfere with the test for nitrate ion.

Test for nitrate ion, NO_3^-:

1. Place a small sample or 1 to 2 mL of solution to be tested in a large test tube.

2. Add 1 to 2 mL of 6 M sodium hydroxide, NaOH, solution. If ammonium ion or ammonia are known to be present, gently boil until red litmus no longer turns blue when inserted into the mouth of the tube. Be very careful not to spill any hot solution on yourself!

3. Add a 2 cm square of aluminum foil to the tube.

4. Push a small wad of cotton one half of the way into the tube to prevent splashing of the solution.

5. Wipe the sides of the mouth of the tube to be sure that no solution is on the walls.

6. Bend a strip of red litmus paper into a V-shape, moisten with distilled water and insert it into the tube so that the tip of the V protrudes into the tube.

7. Gently heat the tube but do not boil. Heat until the aluminum dissolves.

8. Note whether or not the litmus turns blue starting at the tip of the V.

Using the nitrate ion test:

a. Place 1 mL of a 1 M sodium nitrate, $NaNO_3$, solution in a test tube and test for nitrate ion. Record your observations. (J) You should get a positive test for NO_3^-.

b. Place a sample of salt peter about the size of a pencil eraser in a test tube. Test the sample for nitrate ion and record your results.(K)

F. PHOSPHATE ION, PO_4^{3-} A test for phosphate ion in solution is accomplished by adding nitric acid followed by a solution of ammonium molybdate, $(NH_4)_6MoO_{24}$. If phosphate ion is present, a characteristic yellow solid of complex composition is formed.

Test for phosphate ion, PO_4^{3-}:

1. Place a small sample or 1 or 2 mL of the solution to be tested in a test tube.

2. Add 4 or 5 drops of 6 M nitric acid.

3. Add 8 to 10 drops of an ammonium molybdate solution.

4. Gently warm the test tube but do not boil.

5. Note whether or not a yellow solid forms.

Using the phosphate ion test:

a. Place about 1 mL of a 0.1 M sodium phosphate solution, Na_3PO_4, in a test tube and test for phosphate ion. Record your observations. (L) You should get a positive test for PO_4^{3-}.

b. Place a sample of household baking powder about the size of a pencil eraser in a test tube. Dissolve in about 1 mL of water. Test for phosphate ion but be careful of the solution bubbling out of the test tube. Record your observations. If you find phosphate ion, what compound in baking powder contains this ion? (M) Read the label.

c. Place a pencil eraser-sized sample of a household detergent in a test tube and test for phosphate ion. Record your results. (N)

2. TESTING AN UNKNOWN SOLUTION FOR AN ION (OPTIONAL)

Your instructor will issue you a test tube of solution of a compound containing one of the following ions: NH_4^+, Cl^-, CO_3^{2-}, SO_4^{2-}, NO_3^-, or PO_4^{3-}. Record the number of the unknown. (O) Test portions of the solution for each of the ions. Be sure to use clean test tubes rinsed with distilled water. Record the identify of the unknown ion in your solution. (P)

10

REPORT SHEET

Name ________________

Lab Section ________________

Due Date ________________

EXPERIMENT 10

1. Qualitative Tests for Common Ions

A. Ammonia, NH_3, and Ammonium ion, NH_4^+

(A)

(B)

(C)

B. Chloride ion, Cl^-

(D) (E)

C. Carbonate ion, CO_3^{2-}

(F) (G)

D. Sulfate ion, SO_4^{2-}

(H) (I)

E. Nitrate ion, NO_3^-

(J) (K)

F. Phosphate ion, PO_4^{3-}

(L) (M) (N)

2. Testing an Unknown Solution for an Ion (Optional) (P)

10

QUESTIONS

Name ______________________________

Lab Section ______________________

Due Date __________________________

EXPERIMENT 10

1. A compound is dissolved in water and tested for ions. A sample is mixed with a sodium hydroxide solution and litmus paper is held above the mixture. The litmus paper turns blue. Another sample is tested by adding nitric acid and 5 drops of a barium chloride solution. Upon testing, a distinct white solid is formed. What ions are present in the compound and what is the formula of the compound?

2. Give the formulas for ammonia and ammonium ion and explain the differences between the compound and the ion.

3. Give the names and formulas for common polyatomic ions containing:

a. sulfur

b. nitrogen

c. carbon

d. phosphorus

QUESTIONS

Name ______________________

Lab Section ______________________

Results ______________________

10

EXPERIMENT 10

1. A compound is dissolved in water and tested for ions. A sample is mixed with a sodium hydroxide solution and litmus paper is held above the mixture. The litmus paper turns [illegible]. Another sample is tested by adding nitric acid and 5 drops of a [illegible] chloride solution. When tested, a [illegible] white solid is formed. What ions are present in the compound and what is the formula of the compound?

2. Give the formulas for ammonia and ammonium ion and explain the differences between the compound and the ion.

3. Give the names and formulas for common polyatomic ions containing:

a. sulfur

b. nitrogen

c. carbon

d. phosphorus

12 The Chemistry of Oxygen

Objective

The purposes of this experiment are to carry out some chemical reactions involving oxygen and some other elements, to describe observations of the reactions, and to write balanced chemical equations for the reactions.

Discussion

Chemical reactions are processes in which one set of chemicals react to form another set of chemicals. The initial set of chemicals are the reactants, and the chemicals produced are the products. The chemical reactions that an element undergoes with other elements and compounds are characteristic of that element and are called the chemical properties of the element. It is not always possible to know that a chemical reaction has occurred when chemicals are mixed. However, very often a chemical reaction is indicated by some obvious change which accompanies the reaction. Let us consider an example. Sodium is a solid metallic element and chlorine is a green-yellow gas. When these elements are mixed, the white crystalline solid sodium chloride (salt) is formed. The fact that a reaction has occurred is apparent from the change in the appearance of the reactants to the different colored product. The formation of sodium chloride from sodium and chlorine is also accompanied by the release of energy in the form of light and heat. This release of energy also serves as an indication that a reaction has occurred. Other indications that a reaction may have occurred are a color change, the formation of a gas, or the formation of a distinct solid product.

As you carry out the reactions in this experiment, be alert and note any evidence of the occurrence of a chemical reaction. Describe each reaction by noting the nature of the reactants and products and any energy release as heat and light. The reaction can be chemically described by writing a balanced equation. To write an equation, you need to know the formula of the product. For the reactions in this experiment, you will be given the formula of the product. As an example of the description of a reaction, consider the reaction of sodium and chlorine described above:

Solid metal + green gas produce white solid
accompanied by heat and light

The balanced equation for the reaction is: $2Na + Cl_2 \longrightarrow 2NaCl$

Oxygen, which exists in the form of diatomic molecules, O_2, combines with most of the other elements to form binary compounds called oxides. A few examples of oxides are given of the next page:

Na_2O	Sodium oxide
MgO	Magnesium oxide
Al_2O_3	Aluminum oxide
CO_2	Carbon dioxide
N_2O_3	Dinitrogen trioxide
SO_2	Sulfur dioxide
Cl_2O_7	Dichlorine heptoxide

Since oxygen is present in the air, it is possible to form oxides of many elements by heating them in air. As an example of the reaction of oxygen with an element, consider the reaction of calcium. When solid calcium is heated in air, it reacts with oxygen to give the white solid calcium oxide, CaO, or lime. The observations of this reaction are:

Solid metal + colorless gas produce white solid

The balanced equation for the reaction is: $2Ca + O_2 \longrightarrow 2CaO$

Laboratory Procedure

In each of the following you are provided space for notes concerning your observations. On the report sheet summarize your observations and give balanced equations.

1. REACTIONS WITH OXYGEN IN THE AIR

A. ALUMINUM Aluminum metal readily combines with oxygen to form a protective oxide coating that prevents further reaction with oxygen. In fact, one of the characteristics of discarded aluminum is that it forms aluminum oxide very slowly over a period of many years. However, it is possible to combine aluminum and oxygen readily in the presence of a catalyst as illustrated below.

Obtain a 1 cm piece of aluminum wire. Place 3 or 4 drops of a mercury(II) nitrate, $Hg(NO_3)_2$, solution in your crucible. Vigorously rub one end of the wire with steel wool or sand paper to polish it. Dip the polished end of the wire into the solution of mercury(II) nitrate catalyst and place the wire on the edge of the laboratory book with the dipped end protruding over the edge. Rinse out your crucible.

If wire is not available, obtain a small piece of aluminum foil and clean the surface with a piece of steel wool. Place a drop of mercury(II) nitrate solution on the cleaned surface of the foil to serve as a catalyst. Allow the aluminum to sit undisturbed for 5 to 10 minutes. (Continue with the other parts of the experiment and return to this part in 5 or 10 minutes.) (A)

Record any observations and write a balanced equation for any reaction that occurred between aluminum and oxygen. The product is aluminum oxide, Al_2O_3. The mercury(II) nitrate is a catalyst and is not included in the equation.

B. MAGNESIUM Obtain a 2 cm length or magnesium ribbon. Hold the ribbon in your crucible tongs and place it in the Bunsen burner flame. (DANGER: To protect your eyes, do not look directly into the flame when this reaction occurs but turn your head and look out of the corners of your eyes.) Record any observations and write a balanced equation for any reaction that occurred between magnesium and oxygen. The product is magnesium oxide, MgO. (B)

C. COPPER Obtain a small length of copper sheet. Hold the sheet in your crucible tongs or tweezers and place it in the Bunsen burner flame. Move the sheet around in the flame and then remove it and let it cool. Record any observations and write a balanced equation for any reaction that occurred between copper and oxygen. The product is copper(II) oxide, CuO. (C)

D. HYDROGEN Place one or two pieces of mossy zinc in a small test tube. Pour about 1 mL of dilute hydrochloric acid (6 M HCl) into the tube. As the reaction occurs, hydrogen gas will bubble off. Allow the gas to bubble off for a moment and then invert an empty test tube over the top of this tube and fill it with the rising hydrogen gas. Allow the tube to fill with hydrogen for a minute or two. Stopper this tube filled with hydrogen gas with a cork or rubber stopper and then rinse out the original test tube to stop the hydrogen generation. Do not throw the unreacted zinc in the sink. Obtain a wooden splint and ignite it in the Bunsen flame. Remove the stopper from the hydrogen tube and touch the flaming splint to the mouth of the tube. (CAUTION: This is a violent reaction so do not get too close to the tube and, by all means, do not put your face close to the tube.) If you do not observe a reaction, collect another sample and try again. Record any observations and write a balanced equation for any reaction between hydrogen and oxygen. The product is water, H_2O. (D)

2. COMBUSTION

When the term combustion is used in this exercise it refers to the reaction of a carbon-containing compound (organic compound) with oxygen to form carbon dioxide, CO_2, and water, H_2O. The equations for such combustion reactions always include the organic compound and oxygen as reactants and carbon dioxide and water as products.

A. NATURAL GAS Natural gas is a mixture of gases called hydrocarbons; but the main component is methane, CH_4. When you light your Bunsen burner, the combustion of methane occurs producing the expected combustion products as well as heat and light. You may be able to detect the production of water during the combustion of methane by inverting a dry test tube over the edge of the Bunsen burner flame. Hold the test tube with a test tube holder. The water will condense on the sides of the tube. Write a balanced equation for the combustion of methane and describe the reaction. (E)

B. CELLULOSE Cotton fiber consists mainly of the compound cellulose. Place a small wad of cotton about the size of a marble on a ceramic pad or wire gauze. Ignite the cotton with a Bunsen flame. Cellulose is a complex chemical compound which we can represent by the simplest formula, $C_6H_{10}O_5$. Write a balanced equation for the combustion of cellulose and describe the reaction. (F) The small amount of ash comes from impurities in the cotton and some unburned carbon. Do not consider the ash as a product of the reaction.

C. HEXANE Work in a fume hood. Obtain five to ten drops of hexane, C_6H_{14}, in your crucible. Ignite a wooden splint and touch it to the hexane. Record your observations and write a balanced equation for the combustion of hexane. (CAUTION: The crucible may be quite hot.) (G)

D. ISOPROPYL ALCOHOL Work in a fume hood. Obtain five to ten drops of isopropyl alcohol, C_3H_8O, in your cool crucible. (CAUTION: The crucible may be hot from the previous test.) Ignite the alcohol with a flaming splint. Record your observations and write a balanced equation for the combustion of isopropyl alcohol. (H)

3. REACTIONS WITH PURE OXYGEN

A. PREPARATION OF OXYGEN SAMPLES Some pure oxygen can be prepared by decomposing the hydrogen peroxide, H_2O_2, present in a hydrogen peroxide water solution. In this reaction, which is catalyzed by MnO_2, hydrogen peroxide decomposes to form water and oxygen gas. Give the balanced equation for this reaction. (I)

Set up the gas production apparatus shown in Fig. 12-1. You will need a bent and a circular piece of glass tubing. (See Experiment 1, Section 9, for instructions on inserting a tube into a rubber stopper.) Fill three large test tubes with water and stopper them. Invert the tubes into a 400-mL beaker half filled with water. Remove the stoppers from the test tubes. Pour about 20 mL of a 3% hydrogen peroxide solution in the test tube in the gas production apparatus. Then add a sample of solid MnO_2 about the size of a pencil eraser and seal the tube with a stopper attached to rubber tubing. Collect three test tubes of oxygen gas by displacing the water in the inverted test tubes by bubbling the oxygen gas into them. As you fill each tube with gas, remove it and stopper it tightly. If the bubbling of oxygen slows down gently heat the tube with a Bunsen burner flame until the vigorous production of oxygen is apparent. However, DO NOT BOIL the solution. If the hydrogen peroxide solution stops forming oxygen before you fill three tubes just remove the test tube from the apparatus, dump out the liquid and refill the tube with more hydrogen peroxide solution and add some more catalyst. After filling three tubes, rinse out the gas generating tube to stop the reaction.

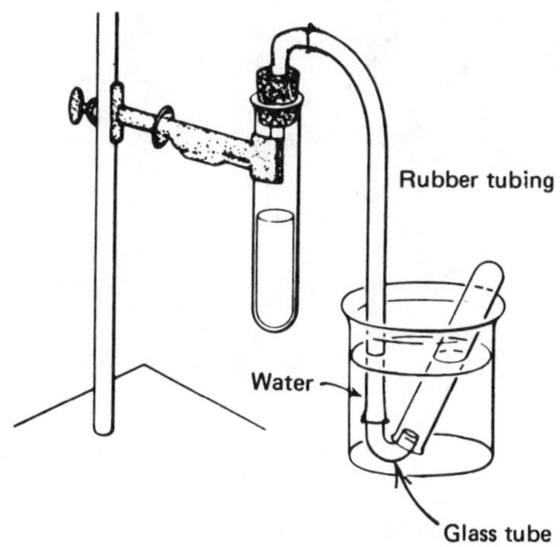

Fig. 12-1 Gas generating apparatus

B. CARBON Obtain a small piece of charcoal (assume it to be carbon, C) and hold it with your tweezers. Heat the charcoal in the Bunsen burner flame. Quickly remove the stopper from a tube of oxygen and immerse the tweezers into the oxygen. Record your observations and write a balanced equation for the reaction between carbon and oxygen to give carbon dioxide, CO_2. The white ash left in the tube is a result of an impurity in the charcoal and is not a product. (J)

C. IRON Obtain a small piece of steel wool and twist it into a long wad about 2 cm long. Hold the wad with your tweezers, heat it in the Bunsen burner flame for a few seconds, and quickly plunge it into a tube of oxygen. Record your observations and write a balanced equation for the reaction between iron and oxygen. Assume that the product is Fe_3O_4. (K)

D. SULFUR Obtain a 2 cm length of glass tubing. Using a test tube rack, carry a test tube of oxygen to the fume hood. Perform the following reaction in a fume hood. Hold the glass tubing on one end with your tweezers using one prong inside and the other outside. Place a sample of sulfur about the size of a wooden match head into the other end of the glass tubing. Carefully heat the sulfur in a Bunsen flame to ignite it. Quickly immerse the glass tubing into the test tube of oxygen. Do not breathe any of the product that is formed. Record your observations and write a balanced equation for the reaction between sulfur and oxygen that produces sulfur dioxide, SO_2. (L)

Fig. 7-1. Gas-generating apparatus

B. CARBON. Obtain a small piece of charcoal (assume it to be carbon) and hold it with your tweezers. Heat the charcoal in the Bunsen burner flame. Quickly remove the stopper from a tube of oxygen and plunge the tweezers into the oxygen. Record your observations and write a balanced equation for the reaction between carbon and oxygen to give carbon dioxide, CO_2. (Assume that any ash left in the tube is a result of an impurity in the charcoal and is not a product.)

C. IRON. Obtain a small piece of steel wool and twist it into a long wad about 2 cm long. Hold the wad with your tweezers, heat it in the Bunsen burner flame for a few seconds, and quickly plunge it into a tube of oxygen. Record your observations and write a balanced equation for the reaction between iron and oxygen. (Assume that the product is iron(III) oxide, Fe_2O_3.)

D. SULFUR. Obtain a 15-cm length of glass tubing. Using a test tube rack, carry a test tube of oxygen to the fume hood. Perform the following operation in the fume hood: Hold the glass tubing at one end with your tweezers using one prong inside and the other outside. Place a sample of sulfur about the size of a wooden match head into the other end of the glass tubing. Carefully heat the sulfur in a Bunsen flame to ignite it. Quickly transfer the glass tubing into the test tube of oxygen. Do not breathe any of the sulfur dioxide that is formed. Record your observations and write a balanced equation for the reaction between sulfur and oxygen that produces sulfur dioxide, $SO_2(g)$.

12

REPORT SHEET

Name ________________

Lab Section ________________

Due Date ________________

EXPERIMENT 12

For any reactions give the observations and equations.

1. Reactions with Oxygen in the Air

(A) a. Aluminum

(B) b. Magnesium

(C) c. Copper

(D) e. Hydrogen

2. Combustion

(E) a. Natural gas

(F) b. Cellulose

(G) c. Hexane

(H) d. Isopropyl alcohol

REPORT SHEET

Name ________________

Lab Section ________________

Due Date ________________

EXPERIMENT 12

3. Reactions with Pure Oxygen

(J) a. Preparation of oxygen samples

(K) b. Carbon

(L) c. Iron

(M) d. Sulfur

12

QUESTIONS

Name ______________________

Lab Section ________________

Due Date ___________________

EXPERIMENT 12

1. List some changes that may serve as indications that a chemical reaction may have occurred.

2. Write completely balanced equations for the following reactions for which the reactants and products are given. Remember that hydrogen, oxygen, nitrogen, fluorine, chlorine, bromine and iodine are among the diatomic elements.

(a) Solid potassium chlorate, $KClO_3$, decomposes upon heating to form solid potassium chloride, KCl, and oxygen gas.

(b) Solid calcium carbonate, $CaCO_3$, decomposes upon heating to form solid calcium oxide, CaO, and carbon dioxide gas, CO_2.

(c) Solid calcium carbide, CaC_2, reacts with liquid water to form solid calcium hydroxide, $Ca(OH)_2$ and acetylene gas, C_2H_2.

(d) Nitrogen dioxide gas reacts with water to give liquid nitric acid, HNO_3, and nitrogen oxide gas.

(e) Solid phosphorus, P, reacts with oxygen gas to form tetraphosphorus decoxide, P_4O_{10}.

(f) Fluorine gas reacts with water to give hydrogen fluoride gas, HF, and oxygen gas.

(g) Solid iron metal reacts with steam, $H_2O(g)$, to given hydrogen gas and Fe_3O_4.

(h) Ammonia gas, NH_3, reacts with oxygen gas to give nitrogen oxide, NO, and water.

(i) Chlorine gas reacts with solid lithium to give solid lithium chloride, LiCl.

(j) Liquid heptane, C_7H_{16}, undergoes combustion with oxygen gasgiving carbon dioxide and water.

(k) Ethane gas, C_2H_6, undergoes combustion with oxygen gas to give carbon dioxide and water.

QUESTIONS

Name ____________________

Lab Section ______________

Due date ________________

EXPERIMENT 12

1. List some observations that serve as indications that a chemical reaction has occurred.

2. Write completely balanced equations for the following reactions in which the reactants and products are given. Remember that hydrogen, oxygen, nitrogen, fluorine, chlorine, bromine and iodine are among the diatomic elements.

(a) Solid potassium chlorate, $KClO_3$, decomposes upon heating to form solid potassium chloride, KCl, and oxygen gas.

(b) Solid calcium carbonate, $CaCO_3$, decomposes upon heating to form solid calcium oxide, CaO, and carbon dioxide gas, CO_2.

(c) Solid calcium carbide, CaC_2, reacts with liquid water to form solid calcium hydroxide, $Ca(OH)_2$, and acetylene gas, C_2H_2.

(d) Nitrogen dioxide gas reacts with water to give liquid nitric acid, HNO_3, and nitrogen oxide gas.

(e) Solid phosphorus, P_4, reacts with oxygen gas to form tetraphosphorus decoxide, P_4O_{10}.

(f) Fluorine gas reacts with water to give hydrogen fluoride gas, HF, and oxygen gas.

(g) Solid iron metal reacts with steam, $H_2O(g)$, to give hydrogen gas and Fe_3O_4.

(h) Ammonia gas, NH_3, reacts with oxygen gas to give nitrogen oxide, NO, and water.

(i) Chlorine gas reacts with solid lithium to give solid lithium chloride, LiCl.

(j) Liquid heptane, C_7H_{16}, undergoes combustion with oxygen gas to give carbon dioxide and water.

(k) Ethane gas, C_2H_6, undergoes combustion with oxygen gas to give carbon dioxide and water.

13 Energy in Chemistry

Objective

The purpose of this experiment is to observe some energy changes that accompany chemical reactions and phase changes.

Discussion

Chemical reactions involve the breaking and forming of chemical bonds. When bonds break and are formed, energy exchanges are involved. Consequently, when a chemical reaction occurs, a corresponding energy exchange takes place. Some chemical reactions release heat to the surroundings and are called exothermic reactions. Other reactions take heat from the surroundings and are called endothermic reactions. Thus, it is sometimes possible to observe the energy exchange involving a reaction by observing whether heat is given off or absorbed. In a chemical reaction that is open to the atmosphere, the energy involved which takes the form of heat is called enthalpy change or heat of reaction. The energy which is involved in a reaction can be shown in the equation to indicate an exothermic or endothermic reaction. For example, an exothermic reaction shows energy as a product

$$CH_4(g) + 2O_2(g) \longrightarrow CO_2(g) + 2H_2O(g) + \text{energy}$$

and an endothermic reaction shows energy as a reactant.

$$\text{energy} + N_2(g) + O_2(g) \longrightarrow 2NO(g)$$

When a chemical dissolves in water, chemical bonds may break and water molecules may be attracted to the various species formed upon dissolving. The energy exchange involved in dissolving is called the enthalpy or heat of solution. The heat of solution can be exothermic or endothermic depending upon the chemical which is dissolved. Is the dissolving process described by the following equation exothermic or endothermic?

$$\text{energy} + NaNO_2(s) \longrightarrow Na^+(aq) + NO_2^-(aq)$$

Some chemical reactions occur spontaneously when the reactants are mixed. However, in many cases, the reactants can be mixed and no reaction occurs. For example, we know that methane can react with oxygen in an exothermic reaction. However, when we turn the gas on in a Bunsen burner and allow the methane to mix with oxygen in the air, nothing happens. It takes a lighted match or spark to start the burning process. Many reactions behave in a similar manner. The reactants can be mixed and no observable reaction occurs. However, with some source of energy, such as a flame or spark, the reaction is initiated. For exothermic reactions the source of energy is just needed to start the reaction and, once it is started, it will proceed on its own and energy will be released. On the other hand, endothermic reactions require a continuous source of energy to occur. An exothermic reaction can warm up the environment and an endothermic reaction can remove heat from the environment.

A catalyst is a chemical that speeds up a reaction without being chemically changed. At a given temperature the catalyzed reaction occurs faster than the uncatalyzed reaction. A catalyst does not affect the amount of heat absorbed or released by a reaction. However, a catalyst does allow the reaction to occur faster. The energy is released or absorbed faster.

Energy exchanges are also involved when chemicals undergo changes of state, such as melting, freezing, boiling, and condensing. Energy is required to melt a solid. The amount of energy required to convert a specific amount of a solid to the liquid state is called the heat of fusion. A pure solid has a unique heat of fusion. When a liquefied solid is cooled, it will usually freeze at the same temperature at which the solid melted. This temperature is the freezing point of the chemical. When freezing occurs, energy is released. The amount of heat released when a specific amount of liquid is solidified is called the heat of crystallization. Since solidification is the reverse of melting, the heat of fusion and heat of crystallization of a chemical will have the same value.

Laboratory Procedure

Space has been provided so that you can record your observations. The observations should be summarized on the report sheet.

1. HEAT OR ENTHALPY OF REACTION

(a) Pour about 1 mL of 6 M hydrochloric acid into a clean test tube. (CAUTION: Acid and base solutions are dangerous so be careful not to spill or splash them.) Pour about 1 mL of 6 M sodium hydroxide solution into another tube. Now, carefully pour the acid into the sodium hydroxide solution. Feel the test tube and record your observations. (A)

The following equation represents the reaction that occurred. Write in the word "energy" on the proper side of the equation to indicate whether energy was released or absorbed.

$$H_3O^+(aq) + OH^-(aq) \longrightarrow 2H_2O$$

(b) Pour about 2 mL of 6 M sodium hydroxide solution into a large test tube. Obtain a 1 cm square of aluminum foil. Place a thermometer in the test tube and observe the temperature. Now drop the foil into the tube and observe the temperature as the reaction takes place. Record your observations below. (C)

The following equation represents the reaction that occurred. Write the word "energy" on the proper side of the equation.

$$6H_2O + 2Al(s) + 2OH^-(aq) \longrightarrow 2Al(OH)_4^-(aq) + 3H_2(g)$$

(c) Pour about 0.5 mL of concentrated (18 M) sulfuric acid in a small test tube. (CAUTION: Concentrated H_2SO_4 is very dangerous. It is more concentrated and dangerous than battery acid) Pour about 5 mL of water in a large test tube. Now, carefully pour the acid into the water, swirl gently and cautiously feel the tube. Record your observations. (E)

The following equation represents the reaction that occurred. Write the word "energy" on the proper side of the equation.

$$H_2SO_4(\ell) + H_2O \longrightarrow H_3O^+(aq) + HSO_4^-(aq)$$

Why should you always add sulfuric acid to water and never add water to sulfuric acid? (G)

(d) Pour about 2 mL of 6 M hydrochloric acid into a small test tube. Obtain a small piece of magnesium metal. Place a thermometer in the test tube and observe the temperature. Now, drop the magnesium in the tube and observe the temperature as the reaction takes place. Record your observations. (H)

The following equation represents the reaction that occurred. Write the word "energy" on the proper side of the equation.

$$2H_3O^+(aq) + Mg(s) \longrightarrow Mg^{2+}(aq) + H_2(g) + 2H_2O$$

2. HEAT OR ENTHALPY OF SOLUTION

(a) Place about 1 g of ammonium chloride, NH_4Cl, in a clean, dry test tube. Place a thermometer in the tube and observe the temperature. Pour in about 5 mL of water. Mix and observe any temperature change. (J)

Write the word "energy" on the proper side of the equation representing the dissolving process.

$$NH_4Cl(s) \longrightarrow NH_4^+(aq) + Cl^-(aq)$$

(b) Pour about 2 mL of water in a large test tube. Obtain about 1 g of calcium chloride. Place a thermometer in the water and observe the temperature. Now, dump in the calcium chloride, then mix and observe the temperature as dissolving occurs. Record your observations below. (L)

Write the word "energy" on the proper side of the equation representing the dissolving process.

$$CaCl_2(s) \longrightarrow Ca^{2+}(aq) + 2Cl^-(aq)$$

30 60 90 120 150 180 210 240 270
58 53 50 48 46 45 44.5 44 44
40 40 40 40 40 40 40 40

300 330 360 390 420 450 480 510 540
44 44 44 44 43.7 43.7 43.5 43.5 43.5

3. CATALYSIS

Hydrogen peroxide can decompose to give oxygen and water.

$$2H_2O_2 \longrightarrow 2H_2O + O_2$$

The reaction is noticeable when the oxygen gas bubbles off. In this section, several chemicals are to be tested for their catalytic effect on the decomposition of hydrogen peroxide. Place seven small test tubes in a test tube rack and put into separate tubes a small amount (5 drops of the solutions or an amount of solid about the size of a match head) of each of the chemicals listed below. Label the tubes or rack to note which test tube has which chemical.

Obtain about 10 mL of a 3% hydrogen peroxide solution. Gently heat the tube in the flame but do not boil. Record any observations. (N)

Pour about 1 mL of the heated hydrogen peroxide solution into each of the tubes in the rack. Observe and describe the catalytic effect or lack of effect for the chemicals in the tubes. Record your results in the following table. (O)

Chemical Tested	Catalytic Effect
0.1 M NaCl	
0.1 M $FeCl_3$	
MnO_2	
Marble ($CaCO_3$)	
Iron	
Aluminum	
Yeast	

Why is it dangerous to store hydrogen peroxide solutions in bottles with iron lids? (P)

4. THE FREEZING POINT OF A COMPOUND

This exercise requires a clock or watch with a second hand. Do the exercise with a partner.

In this section you will observe the freezing of a sample of lauric acid, $C_{12}H_{24}O_2$. To do this, you will measure the temperature of a sample while it cools in a water bath. The temperature will be observed at specific time intervals and the data will consist of a set of time and temperature values. The apparatus needed in the experiment is shown in Fig. 13-1.

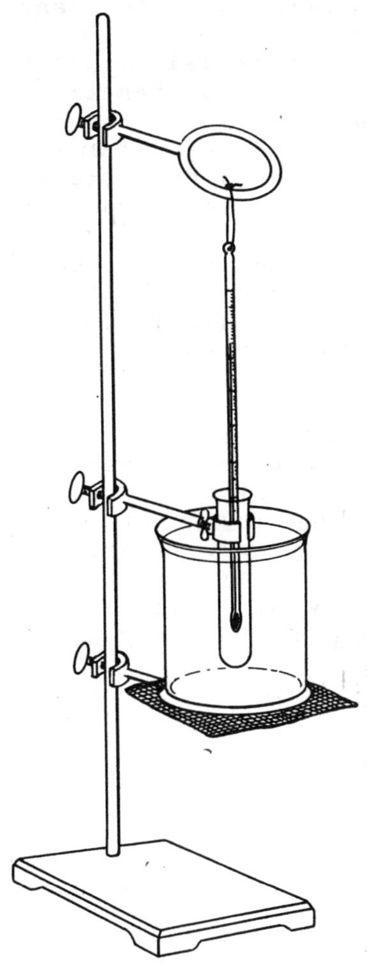

Fig. 13-1

For this part of the experiment you will need a ring stand, two iron rings, a wire gauze, a test tube clamp, a short length of copper wire, two thermometers, a 250-mL beaker and a 400-mL beaker. Set up the apparatus shown in Fig. 13-1 except for the test tube. The beaker can be raised or lowered by adjusting the ring that supports it. Heat a 400-mL beaker half filled with water to about 50 °C or obtain the 50 °C water from the hot water tap. Fill a 250-mL beaker two-thirds full using the 50 °C water and adjust the temperature to about 40 °C by adding cold water. Lower the ring used to support the beaker and place the 250-mL beaker on the gauze supported by the ring.

You will need to work quickly at this time. Obtain a test tube of hot lauric acid and clamp it with the test tube clamp attached to the ring stand. Adjust the clamp so that the suspended thermometer is in the liquid acid and the bulb is about 1 cm from the bottom of the test tube. (See Fig. 13-1.) The thermometer must not touch the sides of the tube. Adjust the ring supporting the beaker of water so that the test tube is submerged in the water. Make sure that as much of the tube is under water as possible. Do not spend too much time getting the tube clamped and immersed into the water. The acid may cool and solidify before you have time to collect any data.

One person watch the clock and record the time and temperature. The other person watch the thermometer and read the temperatures to the nearest 0.1 °C. Read and record the temperature every 30 seconds. The temperature will decrease with time as the acid cools and then level off at some specific value for a few minutes. As soon as the temperature drops below this specific value you may stop collecting data.

When you have finished collecting the data, loosen the copper wire holding the thermometer and remove the test tube with the thermometer in it. Return the tube of solid lauric acid to the designated place. Be careful not to allow any water to enter the test tube of lauric acid.

Using a piece of 8-1/2 by 11 graph paper, plot your data points to obtain a cooling curve. Plot the temperature (8-1/2 side) versus the time (11 side). Connect the plotted points with a smooth curve to form the cooling curve. The freezing point of the lauric acid is the temperature at which the cooling curve levels off for a period of time.

Based on your curve, what is the freezing point of lauric acid? (Q) Describe any evidence that energy is released when solidification or freezing occurred. (R)

14 Stoichiometry

Objective

The purpose of this exercise is to experimentally confirm some principles of stoichiometry.

Discussion

Chemical stoichiometry deals with mass relations in chemical reactions. Stoichiometry is based upon the fact that mass is conserved in a chemical reaction. This means that the total mass of the products in a reaction equals the total mass of the reactants used in the reaction. A balanced equation for a reaction reflects the conservation of mass. The principles of stoichiometry include relating various reactants and products in a reaction through the number of moles of each species involved in a reaction. The coefficients in a balanced equation reveal the relative number of moles of each species involved in a reaction.

In this experiment you will react a sample of sodium hydrogen carbonate with sulfuric acid. The products of the reaction are sodium sulfate, water and carbon dioxide. The sodium sulfate will be isolated by heating to drive off the carbon dioxide and water. An equation for the reaction is

$$2NaHCO_3 + H_2SO_4(aq) \longrightarrow Na_2SO_4 + 2H_2O + 2CO_2$$

Any two species in the reaction can be related by molar ratios revealed by the coefficients. For example, sulfuric acid and sodium hydrogen carbonate are related by the ratios:

$$\left(\frac{2 \text{ moles } NaHCO_3}{1 \text{ mole } H_2SO_4}\right) \quad \text{and} \quad \left(\frac{1 \text{ mole } H_2SO_4}{2 \text{ moles } NaHCO_3}\right)$$

Sulfuric acid and carbon dioxide are related by the ratios:

$$\left(\frac{1 \text{ mole } H_2SO_4}{2 \text{ moles } CO_2}\right) \quad \text{and} \quad \left(\frac{2 \text{ moles } CO_2}{1 \text{ mole } H_2SO_4}\right)$$

Molar ratios are used as factors to determine the number of moles of one species related to a given number of moles of another species. For instance, to determine the number of moles of CO_2 formed when 0.378 moles of H_2SO_4 react, we multiply the moles of H_2SO_4 by the appropriate factor:

$$0.378 \text{ moles } H_2SO_4 \left(\frac{2 \text{ moles } CO_2}{1 \text{ mole } H_2SO_4}\right) = 0.756 \text{ moles } CO_2$$

The number of moles of any substance can be related to the mass using the molar mass as a factor. Using molar masses and molar ratios, a given mass of any species in a reaction can be related to the mass of any other species. For example, suppose we want to determine the number of grams of CO_2 formed when 0.690 g of H_2SO_4 react. Since the equation reveals the relation between the two species on a molar basis, we must work with moles. The molar mass of H_2SO_4 is used to find the number of moles of H_2SO_4 in the given mass. The molar ratio is then used to find the corresponding number of moles of CO_2. Finally, the molar mass of CO_2 is used to find the mass of CO_2 formed. The molar masses have to be deduced as we need them. Consider the steps involved in the calculations. First, the number of moles of H_2SO_4 is found from the mass.

$$0.690 \text{ g} \left(\frac{1 \text{ mole } H_2SO_4}{98.07 \text{ g}}\right)$$

This product is multiplied by the molar ratio relating CO_2 to H_2SO_4 to find the moles of CO_2.

$$0.690 \text{ g} \left(\frac{1 \text{ mole } H_2SO_4}{98.07 \text{ g}}\right)\left(\frac{2 \text{ moles } CO_2}{1 \text{ mole } H_2SO_4}\right)$$

Finally, the molar mass of CO_2 is used to find the grams of CO_2.

$$0.690 \text{ g} \left(\frac{1 \text{ mole } H_2SO_4}{98.07 \text{ g}}\right)\left(\frac{2 \text{ moles } CO_2}{1 \text{ mole } H_2SO_4}\right)\left(\frac{44.01 \text{ g}}{1 \text{ mole } CO_2}\right) = 0.619 \text{ g}$$

In this experiment you will carry out the reaction of sodium hydrogen carbonate and sulfuric acid and collect the sodium sulfate formed. The mass of the sodium sulfate collected will be used in several stoichiometric calculations.

Laboratory Procedure

For the experiment you will need a 100-mL beaker, a 250-mL beaker, an evaporating dish and a watch glass cover. Obtain a test tube of $NaHCO_3$ and pour the contents into the 100-mL beaker. Record the number of the sample. (A) If any of the sodium hydrogen carbonate remains in the tube rinse it into the beaker with a small amount of distilled water. Measure out 7 mL of 1 M sulfuric acid in a graduated cylinder. Slowly pour the sulfuric acid into the beaker of sodium hydrogen carbonate. Be careful not to let the solution bubble out of the beaker as you add the acid. After you have added all of the acid and the bubbling has stopped, rinse the sides of the beaker with a small amount of distilled water using your wash bottle. Be sure to use a small amount of water.

Add two or three drops of methyl orange indicator to the beaker. Obtain a dropper bottle of sulfuric acid and add the acid to the beaker drop by drop until the color changes from yellow to orange. It is very important not to add too much acid at this point. If you think that you added too much acid or if the solution turns red rather than orange you will need to start the experiment over.

Place two boiling chips in an evaporating dish, put the watch glass cover on the dish and weigh this combination to 0.01 g. (C) Pour the solution from the 100-mL beaker into the evaporating dish and cover with the watch glass. Place the covered dish on top of a 250-mL beaker supported by a ring stand as shown in Figure 14-1. The beaker will serve as an air bath to heat the dish. Adjust the height of the ring so that the hot part of the Bunsen flame will reach the bottom of the beaker. Light the Bunsen burner and heat the beaker with the full Bunsen flame. Allow the solution in the dish to gently boil until all of the water has boiled off. The heating may take some time. Continue heating until all of the water has evaporated, even the small drops which condense on the underside of the watch glass. IMPORTANT NOTE: If white fumes or smoke evolve from the dish near the end of the heating, stop the experiment and consult your instructor. After heating the dish, allow it to cool to room temperature. When it is cool enough to touch you can move the dish to the top of the desk so that it will cool faster. The dish must be cooled to room temperature before you weigh it. When it is cool, weigh the covered dish to 0.01 g. (B)

Use your experimental data to determine the mass of the sodium sulfate, Na_2SO_4, that was deposited in the dish. (D) Using the mass of sodium sulfate and stoichiometric factors, calculate the number of grams of sodium hydrogen carbonate in your original sample (E), the number of grams of sulfuric acid used in the reaction (F), and the number of moles of carbon dioxide produced in the reaction (G).

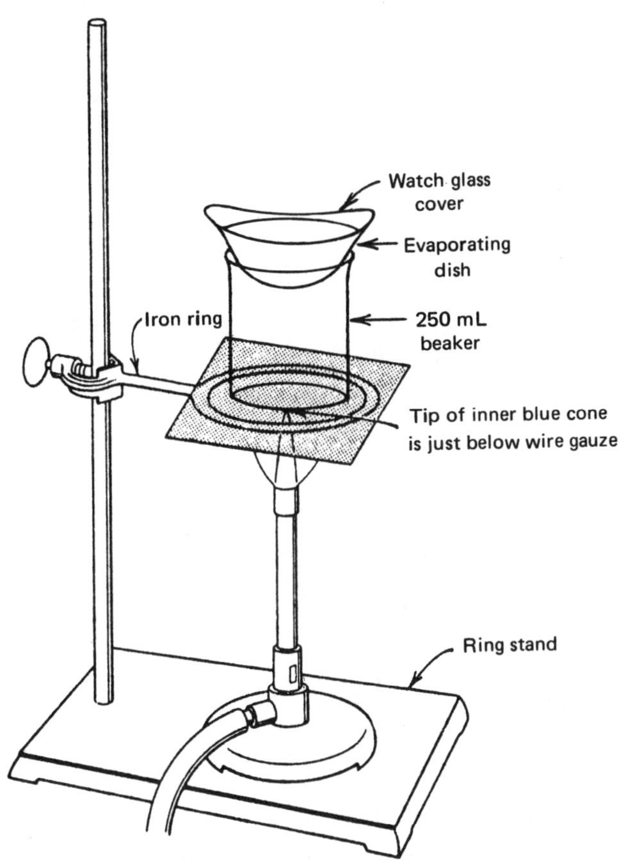

Fig. 14-1 Apparatus for Air Bath.

Place 4 ceramic boiling chips in an evaporating dish, put the watch glass cover on the dish and weigh this combination to 0.01 g. (C) Pour the solution from the 100-mL beaker into the evaporating dish and cover with the watch glass. Place the covered dish on top of a 250-mL beaker supported by a ring stand as shown in Figure 14-1. The beaker will serve as an air bath to heat the dish. Adjust the height of the ring so that the hot part of the Bunsen flame will reach the bottom of the beaker. Light the Bunsen burner and heat the beaker with the full Bunsen flame. Allow the solution in the dish to gently boil until all of the water has boiled off. The boiling may take some time. Continue heating until all of the water has evaporated even the small drops which condense on the underside of the watch glass. IMPORTANT NOTE: If water is not entirely evolved from the dish near the end of the period, save the experiment and consult your instructor. After heating the dish, allow it to cool to room temperature. When it is cool enough to touch you can move the dish to the top of the desk to let it cool faster. The dish must be cooled to room temperature before you weigh it. When it is cool, weigh the covered dish to 0.01 g. (E)

Use your experimental data to determine the mass of the sodium sulfate, Na_2SO_4, that was formed in the dish. (F) Using the mass of sodium sulfate and stoichiometric factors, calculate the number of grams of sodium hydrogen carbonate in your original sample (G), the number of grams of sulfuric acid used in the reaction (H), and the number of moles of carbon dioxide produced in the reaction (I).

Fig. 14-1 Apparatus for Air Bath

14

REPORT SHEET

Name _______________

Lab Section _______________

Due Date _______________

EXPERIMENT 14

(A) Sample Number _____________

(B) Mass of dish, cover and Na_2SO_4 _______________

(C) Mass of dish and cover _______________

(D) Mass of sodium sulfate _______________

(E) Mass of sodium hydrogen carbonate. (Give setup and calculated answer.)

(F) Mass of sulfuric acid. (Give setup and calculated answer.)

(G) Moles of carbon dioxide. (Give setup and calculated answer.)

14

QUESTIONS

Name ________________________

Lab Section _________________

Due Date ____________________

EXPERIMENT 14

1. The compound calcium carbide, CaC_2, is made by reacting calcium carbonate with carbon at high temperatures. The unbalanced equation for the reaction is:

$$CaCO_3 + C \longrightarrow CaC_2 + CO_2$$

(a) Balance the equation.

(b) How many grams of $CaCO_3$ are needed to form 346 g of CaC_2?

(c) How many grams of C are needed to react with 983 g of $CaCO_3$?

2. The fertilizer ammonium sulfate can be made by the following set of reactions:

$$N_2 + 3H_2 \longrightarrow 2NH_3$$

$$2NH_3 + H_2SO_4 \longrightarrow (NH_4)_2SO_4$$

(a) How many grams of ammonium sulfate can be formed from 525 g of H_2?

(b) How many grams of nitrogen are needed to form 898 g of ammonium sulfate?

15 Stoichiometry Worksheet

Name ________________

15

Lab Section ________________

Due Date ________________

1. Ammonia gas reacts with oxygen gas according to the following equation:

$$4\ NH_3 \quad + \quad 5\ O_2 \longrightarrow 4\ NO \quad + \quad 6\ H_2O$$

a. How many moles of oxygen are needed to react with 49 moles of ammonia?

b. How many grams of ammonia are needed to react with 4.35 moles of oxygen?

c. How many grams of oxygen are needed to react with 2.78×10^4 g of ammonia?

2. Metallic titanium is manufactured by the following reaction:

$$TiCl_4 + 4\ Na \longrightarrow 4\ NaCl + Ti$$

a. How many moles of Na are needed to form 68 moles of Ti

b. How many grams of $TiCl_4$ are needed to form 5.67×10^3 g of Ti?

c. How many grams of Ti are formed when 362 g of $TiCl_4$ and 184 g of Na react?

3. Elemental antimony is made by the following set of reactions:

$$Sb_2S_3 + 5\ O_2 \longrightarrow Sb_2O_4 + 3\ SO_2$$

$$Sb_2O_4 + 4\ C \longrightarrow 2\ Sb + 4\ CO$$

How many grams of O_2 are needed to prepare 1.00 g of Sb?

16 Gas Laws

Objective

The purposes of this experiment are to:

1. Measure the volume of a gas sample at various pressures and a constant temperature.

2. Compare the measured volume of a gas with the volume calculated using Boyle's Law.

3. Measure the volume of a gas sample at various temperatures and a constant pressure.

4. Compare the measured volume of a gas with the volume calculated using Charles' Law.

5. Plot a graph of the volume of a gas sample versus the temperature at a constant pressure and to determine a value of "absolute zero" from the graph.

Discussion

The experiment is designed to allow you to confirm Boyle's Law and Charles' Law. Boyle's Law states the volume of a gas sample at a constant temperature is inversely proportional to the pressure. A mathematical expression of the law is $V \alpha 1/P$ or as an algebraic equation $V = k/P$ where k is a proportionality constant. Boyle's Law means that if the pressure of a gas sample is increased, the volume will decrease and if the pressure is decreased, the volume will increase. This is what is meant by inversely proportional. Because of this proportionality, we can calculate the new volume of a gas sample if we know the initial volume and pressure and the final pressure of the gas. The new volume of a gas sample, V, which results when the pressure of the sample is changed is found by multiplying the initial volume, V_i, by a ratio of the pressure corresponding to the change.

$$V = V_i \left(\frac{P_i}{P_f} \right)$$

The pressure ratio should be greater than one if the initial pressure is decreased. This will give a new volume that is greater than the initial volume. The pressure ratio should be less than one if the initial pressure is increased. This will give a new volume that is less than the initial volume.

In this experiment you will measure the pressure and volume of a gas sample using a simple apparatus. It consists of an air sample trapped between the sealed end of a narrow glass tube and a "plug" of mercury that is free to move up and down the tubing. Figure 16-1 gives an illustration of the apparatus. It is very important that you handle this tube with the utmost care. Handle it gently and do not shake it or move it abruptly. If at any time the mercury plug separates into parts or if any mercury comes out of the tube notify your instructor. The tube is a delicate piece of equipment that will serve you well if you handle it carefully.

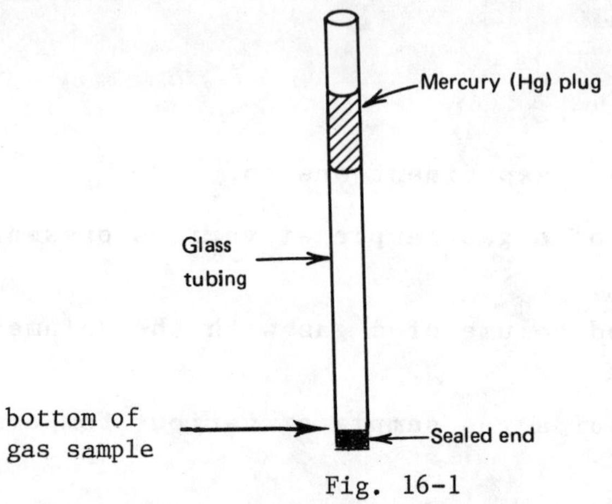

Fig. 16-1

If the apparatus is in a horizontal position, the plug will adjust to a position in which the pressure of the gas in the tube equals the atmospheric pressure. That is, as shown in Figure 16-2, the pressures at points A, B and C will be equal to the atmospheric pressure.

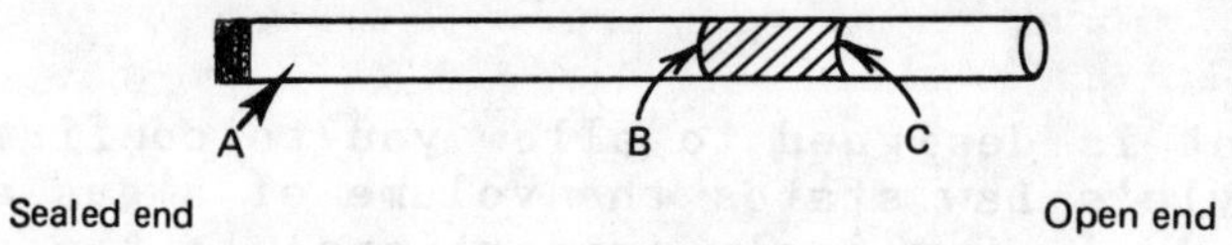

Fig. 16-2

When the tube is held vertically as shown in Figure 16-3, the plug will adjust to a position in which the pressure of the gas balances the pressure of the atmosphere plus the additional pressure exerted by the mercury plug. That is, the pressure of the gas equals the atmospheric pressure plus the pressure exerted by the column of mercury corresponding to the height of the mercury plug.

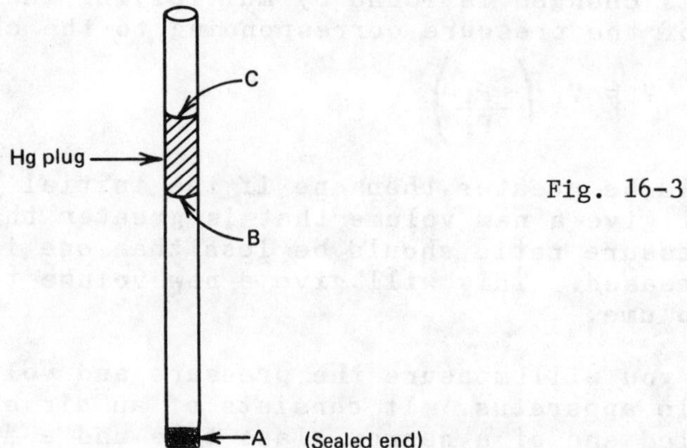

Fig. 16-3

For example, suppose the atmospheric pressure is 765 torr and the height of the mercury column from point B to point C in Figure 16-3 is 16 mm of mercury. The pressure of the gas will be: 765 torr + 16 torr = 781 torr.

When the tube is held with the open end pointing down, as shown in Figure 16-4, the pressure of the trapped gas is lower than the atmospheric pressure by the pressure corresponding to the height of the mercury plug. That is, the pressure of the gas equals the atmospheric pressure minus the pressure corresponding to the height of the column of mercury in the plug.

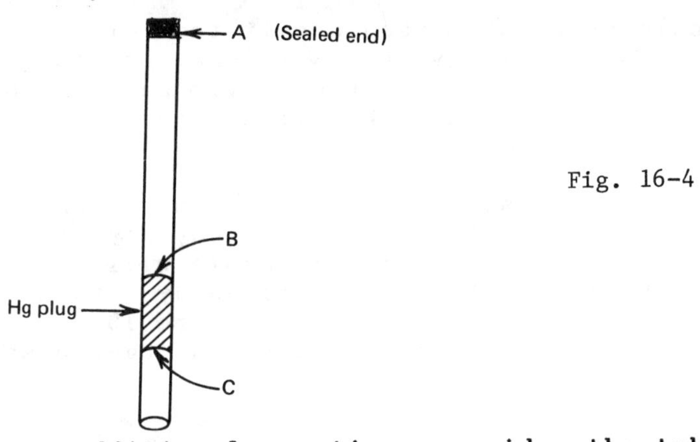

Fig. 16-4

As additional practice, consider the tube shown in Figure 16-5. In the figure, the tube is kept in a container at reduced pressure. The pressure in the container is 450 torr and the height of the mercury column from point B to point C is 16 mm of mercury. The pressure inside the tube is greater than the pressure inside the container by an amount equal to the 16 torr column. What is the pressure of the gas in the tube? _________

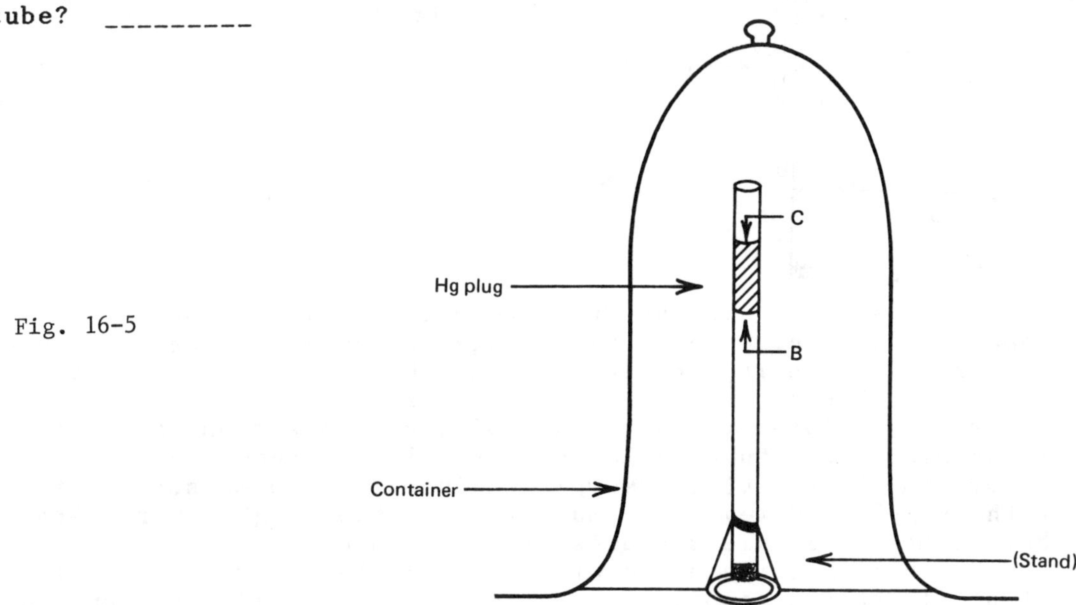

Fig. 16-5

Charles' Law states that the volume of a gas sample at a constant pressure is directly proportional to the temperature. A mathematical expression of the law is $V \alpha T$ or as an algebraic equation $V = kT$ where k is a proportionality constant. Charles' Law means that if the temperature of a gas sample is increased, the volume will increase and if the temperature is decreased, the volume will decrease. This is what is meant by directly proportional. Because of this proportionality, we can calculate the new volume of a gas sample if we know the initial volume and temperature and the final temperature of the gas. The new volume of the gas sample, V, which results when the temperature of the sample is changes is found by multiplying the initial volume, V_i, by a ratio of the temperatures corresponding to the change.

$$V = V_i \left(\frac{T_f}{T_i} \right)$$

The temperature ratio should be greater than one if the initial temperature is increased. This will give a new volume that is greater than the initial volume. The temperature ratio should be less than one if the initial temperature is decreased. This will give a new volume that is less than the initial volume.

In this experiment you will measure the volume and temperature of a gas sample using an apparatus like that used to study Boyle's Law. (See Figure 16-1.) The volume of a trapped gas sample will be measured at various temperatures. The temperature is measured by attaching the mercury-plugged tube to a thermometer as shown in Figure 16-6. A short length of rubber tubing is used to hold the tube to the thermometer.

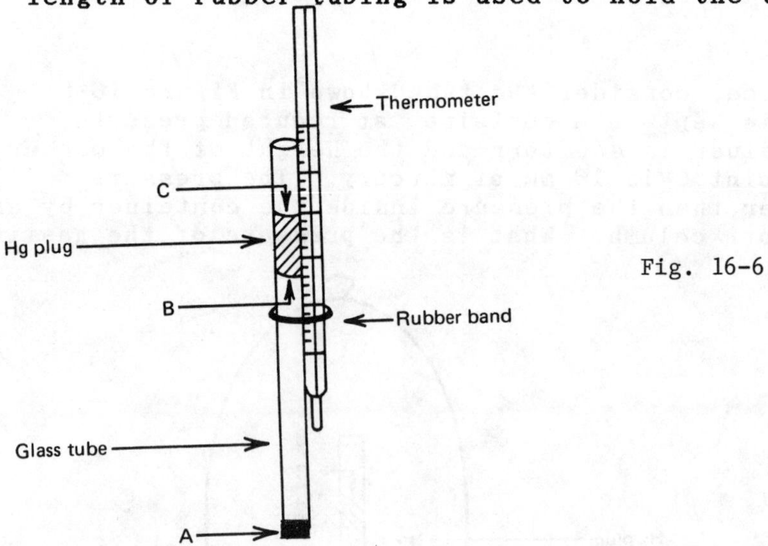

Fig. 16-6

The gas sample in the tube has a cylindrical shape because of the shape of the tube. The volume of the gas sample is the volume of the cylindrical space it occupies in the tube. It is not possible for us to measure this volume directly. However, the volume of the gas sample is directly related to the height of the sample from the bottom of the gas sample in the tube to the mercury plug. Consequently, we can measure the height of the sample and use it as a measure of the volume of the sample. However, we must measure the heights very carefully so that we have reliable measures of the volumes.

In this experiment you will study Boyle's Law by measuring the height of the gas sample at various pressures while the temperature is constant. Since the height of the gas sample is related to the volume of the sample, the height can be used as a measure of the volume. You will study Charles' Law by measuring the height of the gas sample at various temperatures while the pressure is constant. Again the height will serve as a measure of the volume of the gas sample.

Laboratory Procedure

A Reminder: The gas law apparatus that you will use for this experiment is very fragile and must be handled with care. Do not shake it or handle it roughly. If the mercury plug separates, consult your instructor. If you break the tube and spill any mercury, notify your instructor immediately. Do not try to clean up the mercury but be sure that any spills are reported so that they can be completely cleaned up.

You will make your height measurements with a metric rule. It is important to make these measurements as carefully as possible and try to estimate to the nearest 0.1 mm or so.

Obtain a gas law apparatus, thermometer, metric ruler and a large test tube. You will also need a sharp pencil or fine-tip pen.

1. BOYLE'S LAW

Record the barometric pressure as provided. (A) Use this as the initial pressure. Obtain a clean sheet of 8-1/2 by 11 white paper. Fold it in half (8-1/2 by 5-1/2) and fold again (8-1/2 by 2-/1/4). Lay the tube with the mercury plug flat on the paper and, using the pencil or a fine tipped pen, carefully mark the paper with points corresponding to both ends of the plug. It is important to make very thin lines. Using the ruler, measure the distance between the marks as closely as possible. Record this as the height of the mercury column. (B) Lay the tube flat on the paper and align the sealed end of the tube with the edge of the paper. Make sure that the tube is straight along the paper and the end is flush with the edge of the paper. Use the pencil to carefully mark the position of the bottom of the mercury plug and the bottom of the gas sample. Measure the distance between the marks and record this as the initial volume. (C)

Hold the tube flat on the paper and lift the tube and paper so that the open end of the tube is pointing straight up. Be sure that the sealed end of the tube is aligned with the bottom of the paper and that the tube is in a vertical position. Carefully mark the position of the bottom of the mercury plug and the bottom of the gas sample. Measure the distance between the marks and record this as the volume. (D) The pressure of the sample is the atmospheric pressure plus the pressure corresponding to the plug height. Calculate and record this pressure. (E)

Hold the tube flat on the paper and carefully lift the tube and paper so that the open end of the tube is pointing straight down. Be sure that the sealed end of the tube is aligned with the edge of the paper and that the tube is vertical. Carefully mark the position of the bottom of the mercury plug and the end of the gas sample. Measure the distance between the marks and record this as the volume. (F) The pressure of the sample is the atmospheric pressure minus the pressure corresponding to the plug height. Calculate and record this pressure. (G)

To confirm Boyle's Law we will use the law to calculate the expected volumes and compare these to the measured volumes. Remember that we will use height of the gas sample as an expression of volume. Record your calculated results in the data table. Use the atmospheric pressure as the initial pressure and height (C) as the initial volume. Calculate the volume of the gas when the pressure is changed to (E). (H) Compare the calculated result with the experimental volume (D) by determining the percent error. The percent error is found by subtracting the smaller value from the larger value, dividing by the calculated value and multiplying by 100. (I)

Calculate the volume of the gas when the pressure is changed to (G). (J) Determine the percent error between this value and the measured height value (F). Record the percent error. (K)

2. CHARLES' LAW

Tape a strip of magic tape along the length of the large test tube. Make sure that the tape is straight and flat against the tube. Attach the tube to a clamp supported by a ring stand. Make sure that the tape is not obscured by the clamp. Attach the gas tube to a thermometer as shown in Figure 16-6.

Samples of water at various temperatures are to be placed in the test tube. The thermometer and gas tube will be immersed in the water. By holding the gas tube against the side of the test tube, the positions of the bottom of the gas sample and the bottom edge of the mercury plug can be marked on the magic tape. It is very important to make the marks on the tape as carefully as possible and to make very narrow lines with a sharp pencil or fine tipped pen. The temperature can be read from the thermometer. Take care not to allow any water to enter the gas tube.

Fill the test tube with enough room-temperature water so that the thermometer can be placed in it to cover the gas sample with water. Hold the thermometer and tube in the water for a few minutes. Mark the positions of the bottom of the gas sample and the bottom of the mercury plug. Read and record the temperature. (L) Empty the water from the tube.

Bring a small beaker of water to the boil. Using a length of paper towel to hold the beaker, transfer some of the boiling water to the test tube. Hold the the thermometer and gas tube above the hot water for a minute and then slowly immerse it into the hot water. If you suddenly plunge the thermometer and gas tube into the hot water the expansion of the gas may separate the mercury plug. Immerse the thermometer and tube into the hot water slowly. After a minute or so line up the bottom of the gas sample with the mark on the tape and mark the position of the bottom of the mercury plug. Read and record the temperature. (N)

Add some cold water to the test tube so that the temperature is around 40 °C. Stir the water so that the temperature is uniform. Wait a minute or so and then line up the bottom of the gas sample with the mark on the tape. Mark the position of the bottom of the mercury plug. Read and record the temperature. (P) Empty the water from the tube.

Add some crushed ice to the test tube and some water. Stir to mix. Add enough ice so that you have an ice and water mixture. Hold the thermometer and gas tube in the ice water. Wait about two minutes, line up the tube and mark the position of the bottom of the mercury plug. Read and record the temperature. (R)

Remove the tube from the clamp and dump out the water. Using your ruler, carefully measure the marks corresponding to the various heights. These are height at room temperature (M), height in hot water (O), height in about 40 °C water (Q), and height in ice water (S). These heights will be used as expressions of volumes for the gas sample.

Use the volume in room temperature water as the initial volume and calculate the theoretical volume at the temperature of ice water using Charles' Law. Remember that you need to use Kelvin temperatures in Charles' Law calculations. (T) Compare this with your experimental value for ice water and calculate the percent error. (U)

By graphing your experimental data it is possible for you to estimate the Celsius temperature that would correspond to a gas volume of zero. This temperature is called absolute zero and corresponds to zero on the Kelvin temperature scale. See appendix 3 for a review of graphing techniques. Plot the height of the gas sample versus the Celsius temperatures on the graph paper provided. Note that the graph paper has the temperature axis already numbered. You need to number the height axis. When deducing the range of your height data the range should extend from zero height to your largest height value. Once you have plotted the points draw the best straight line through the plotted points and, using a ruler, extrapolate the line to the Celsius temperature corresponding to zero volume which will be zero height using your data. Record this temperature as the Celsius equivalent of absolute zero. (V)

16

REPORT SHEET

Name _______________

Lab Section _______________

Due Date _______________

EXPERIMENT 16

1. Boyle's Law

(A) initial pressure ____________ (B) mercury plug height _______

(C) initial volume ______________

(D) volume ______________________ (E) pressure _________________

(F) volume ______________________ (G) pressure _________________

(H) calculated volume ___________ (I) percent error ____________

(J) calculated volume ___________ (K) percent error ____________

2. Charles' Law

(L) temperature _________________ (M) volume ___________________

(N) temperature _________________ (O) volume ___________________

(P) temperature _________________ (Q) volume ___________________

(R) temperature _________________ (S) volume ___________________

(T) calculated volume ___________ (U) percent error ____________

(V) experimental absolute zero ____________

Plot your graph on the other side of this paper.

Volume–Temperature Relationship of a Gas

mm

0

−300 −200 −100 0 100 200

Temperature, °C

16

QUESTIONS

Name ______________________________

Lab Section ______________________

Due Date _________________________

EXPERIMENT 16

1. Explain why we were able to use the height of the column of gas as a measure of the volume of gas in the gas tube apparatus.

2. In the Boyle's Law experiment, why were we able to assume that the temperature was constant?

3. In the Charles' Law experiment, why were we able to assume that the pressure was constant?

4. If the actual value of absolute zero is −273 °C, calculate the percent error between your value and the actual value of absolute zero.

17 The Molar Mass of a Gas

Objective

The purpose of this experiment is to determine the number of grams per mole of a gas by measuring the pressure, volume, temperature and mass of a sample.

Discussion

The molar mass of a gaseous compound can be determined by experiment even though the formula or composition of the compound are not known. In other words, it is possible to find the molar mass of a gas even if we do not know its identity. The molar mass can be determined by using the fact that the number of moles of a gas sample can be related to the pressure, volume and temperature by the gas laws. The mass, pressure, volume and temperature of a gas sample are measured experimentally and used to calculate the molar mass. There are two methods which can be used to calculate the molar mass of a gas from experimental data.

To determine the molar mass of a gaseous substance from the mass, volume, temperature, and pressure of a sample the number of moles in the sample is first calculated. One way to do this is to convert the volume to standard temperature and pressure and use the molar volume to find the moles. The mass of the sample is then divided by the number of moles to give the molar mass. As an example, consider a 0.508 g sample of a substance that occupies 522 mL volume at 100 °C and 0.960 atmospheres. (Fill in the following blanks.) First, the volume at STP is found.

$$\begin{array}{lcl} 0.522 \text{ L} & \rightarrow & V \\ 373 \text{ K} & \rightarrow & 273 \text{ K} \\ 0.960 \text{ atm} & \rightarrow & 1.00 \text{ atm} \end{array}$$

$$0.522 \text{ L} \left(\text{---------}\right)\left(\text{----------}\right)$$

Next, the molar volume is used to find the number of moles.

$$0.522\text{L} \left(\text{--------}\right)\left(\text{--------}\right)\left(\frac{1 \text{ mole}}{22.4 \text{ L}}\right) = \text{--------} \text{ moles}$$

The molar mass is found by dividing the mass of the sample by the number of moles. Fill in the blanks on the next page.

$$\left(\frac{0.508\text{ g}}{\_\_\_\_\_\_\_\_\_\text{ moles}}\right) = \frac{\_\_\_\_\_\_\_\_\_\text{ g}}{1\text{ mole}}$$

Another way to find the molar mass of a gas involves the ideal gas law, PV = nRT. Recall that n represents the number of moles. The number of moles of a gas can be found from the pressure, volume and temperature of a sample: n = PV/RT

As an example, consider a 0.508 g sample of a substance that occupies 522 mL volume at 100 °C and 0.960 atmospheres. (Fill in the following blanks.)

$$n = \left(\frac{\_\_\_\_\_\_\_\_\_\_\text{ atm} \quad \_\_\_\_\_\_\_\_\_\_\_\text{ L}}{(0.0821\text{ L atm/K mole}) \quad \_\_\_\_\_\_\_\_\_\_\_\text{ K}}\right)$$

The molar mass is found by dividing the mass of the sample by the number of moles.

$$\left(\frac{0.508\text{ g}}{\_\_\_\_\_\_\_\_\_\text{ moles}}\right) = \frac{\_\_\_\_\_\_\_\_\_\text{ g}}{1\text{ mole}}$$

In this experiment a sample of gas will be collected by water displacement. That is, the gas will be bubbled into a container of water and as the gas accumulates, it will displace the water. The volume of the sample is the volume the gas occupies in the container. The temperature can be found by measuring the temperature of the water in contact with the gas. The mass of the gas sample is found by weighing a small tank or cylinder of gas, delivering the sample and reweighing the cylinder.

The pressure of the sample can be found by making sure that the pressure of the gas sample is the same as the atmospheric pressure. The atmospheric pressure can be measured with a barometer. However, when a gas is collected by water displacement, it becomes saturated with water vapor. This means that once the gas sample is collected, it will be a mixture of the gas and water vapor. This does not affect the volume or temperature of the gas but the measured pressure is the pressure of the mixture. That is, the measured pressure is the total pressure of the gas and the water vapor. To determine the pressure of the gas sample, the pressure of the water vapor can be subtracted from the total pressure. Appendix 1 lists the vapor pressure of water at various temperatures. To obtain the pressure of the gas sample in the experiment, look up the vapor pressure of water at the measured temperature and subtract this pressure from the measured atmospheric pressure.

Laboratory Procedure

The glassware needed for this experiment is a 800-mL beaker and a 250-mL Erlenmeyer flask. Place a piece of tape or a length of gummed label along the neck of the flask near the top. Fill the flask to the brim with tap water and put about 300 mL of water into the beaker. Allow the water to stand for a while to be sure that it is at room temperature. Set up the apparatus as shown in Figure 17-1. Fill the beaker with about 300 mL of water. Stopper the flask with a rubber stopper and do not allow any air to become trapped in the flask under the stopper. Invert the flask through the iron ring which will hold it in position. Lower the ring with the flask so that the mouth of the flask is below the water level in the beaker. Secure the ring to support the flask. Use a spatula to remove the rubber stopper from the mouth of the flask. The stopper can remain in the beaker since it will not interfere with the experiment.

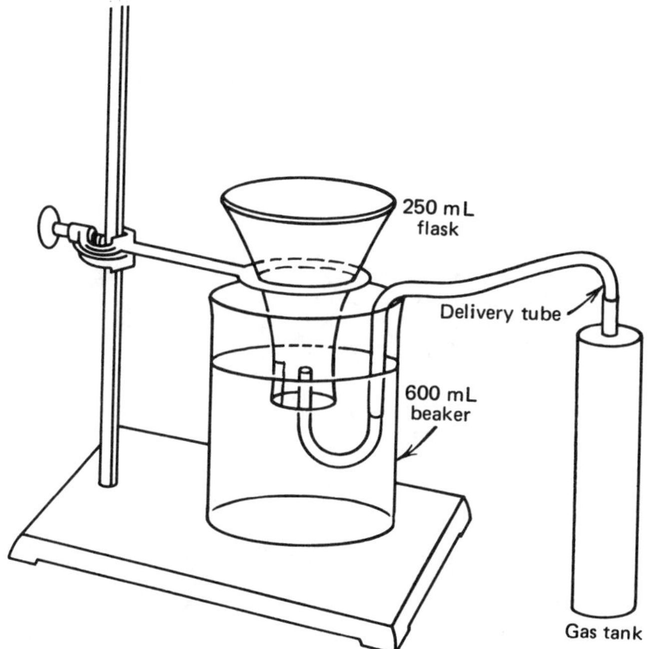

Fig. 17-1 Apparatus for collecting gas sample.

Obtain a small tank of gas. Use a paper towel to gently wipe the tank to make sure that it is clean. Weigh the tank to 0.01 g. (A) Attach a plastic delivery tube to the valve on the tank and insert the glass tube on the other end into the mouth of the flask in the beaker of water. Push the valve on the tank to deliver gas to the flask. Continue to deliver gas until the flask is nearly full of gas. That is, deliver gas until the gas level is near the neck of the flask. Do not allow any gas to escape from the flask. If any gas escapes you will have to start the experiment over. Carefully remove the delivery tube from the gas tank, gently clean and dry the tank with a paper towel and weigh the tank to 0.01 g. (B) Place a thermometer in the beaker.

Move the flask of gas so that the level of water in the flask is the same as the water level in the beaker. This will adjust the pressure in the flask to atmospheric pressure. Use a pen or pencil to place a mark on the tape to indicate the water level. Remove the flask from the water and fill it with water to the level corresponding to the mark on the tape. This volume of water will correspond to the volume of the gas sample. Carefully pour the water from the flask into a graduated cylinder and measure the volume to the nearest 1 mL. Record this as the volume of the gas sample. (C).

Record the temperature of the water in the beaker. (D) Assume that the temperature of the gas sample is the same as the water. Record the atmospheric pressure (E) and look up the vapor pressure of water at the temperature of the gas sample in Appendix 1. (F)

Repeat the experiment to obtain a second set of data. Calculate the molar mass of the gas using each set of data. If the results are within 10 percent of one another, express your answer as the average of the two results. If your two results seem to be too far apart consult your instructor. (G)

17

REPORT SHEET

Name ________________

Lab Section ________________

Due Date ________________

EXPERIMENT 17

(A)	Mass of tank	__________	__________	__________
(B)	Mass of tank - sample	__________	__________	__________
	Mass of sample (A) - (B)	__________	__________	__________
(C)	Graduated cylinder volume	__________	__________	__________
(D)	Temperature of sample	__________	__________	__________
(E)	Atmospheric pressure	__________	__________	__________
(F)	Vapor pressure of water	__________	__________	__________
	Pressure of sample (E) - (F)	__________	__________	__________

Summary of Data

Mass of sample	__________	__________	__________
Volume in liters	__________	__________	__________
Temperature in Kelvin	__________	__________	__________
Pressure in atmospheres	__________	__________	__________

(g) Give the setup and results of molar mass calculations below.

Average Molar Mass (G) ______________________

17

QUESTIONS

Name ________________________

Lab Section _________________

Due Date ____________________

EXPERIMENT 17

1. Referring to the experimental determination of the molar mass, explain how and why each of the following factors would affect your calculated result. That is, would the calculated result be greater than it should be, less than it should be, or not affected.

(a) The measured temperature is lower than the actual temperature of the gas.

(b) The measured volume is higher than the actual volume of the gas.

(c) Some of the gas sample escapes from the flask before the volume and pressure are measured.

2. The molar mass of a gas is determined by collecting a gas sample by water displacement.

(a) Using the following data, calculate the molar mass of the gas: Sample volume, 144 mL; Sample temperature, 23 °C; Sample mass, 0.247 g; Atmospheric pressure, 0.992 atm.

(b) If the gas in part (a) contains 85.5% C and 14.5% H, determine the empirical formula and use the result of part (a) to determine the actual formula.

QUESTIONS

Name ____________________

Section ____________________

Instructor ____________________

EXPERIMENT [illegible]

1. [illegible] experiment [illegible] the [illegible] what effect [illegible] have on the calculated molar mass [illegible]

(a) The measured temperature is lower than the actual temperature [illegible]

(b) The measured volume is higher than the actual volume of the gas.

(c) Some of the [illegible] sample [illegible] on the flask before the volume and [illegible] are measured.

2. The molar mass of [illegible] can be determined by [illegible] the sample by [illegible]

(a) Using the following data, calculate the molar mass of the [illegible]: [illegible] volume, 114 mL; [illegible] temperature, [illegible]; Sample mass, [illegible] g; Atmospheric pressure, 0.982 atm.

[illegible] the [illegible] contains [illegible] and [illegible] determine the [illegible] formula and use the result of part (a) to determine the [illegible] formula.

18 Gas Laws Worksheet

Name ________________

Lab Section ________________

Due Date ________________

18

1. A sample of a gas occupies a volume of 335 mL at 22 °C and 749 Torr. What volume will it occupy at 22 °C and 1.54 atm?

2. A sample of a gas occupies a volume of 42.5 mL at 20 °C and 1 atm. What volume will it occupy at 357 K and 1 atm?

3. A gas sample is stored in a steel tank with a thermometer and a pressure gauge attached. Initially, the temperature reads 21 °C and the pressure gauge reads 1.46 atm. What will the pressure of the gas sample be when the gas is cooled to -40 °C?

4. A gas sample is stored in a flexible container and occupies a volume of 275 L at STP (Standard Temperature and Pressure is 0 °C and 1.00 atm.) What volume with the gas occupy at 798 Torr and 259 K?

5. A gas sample occupies a volume of 23.5 L at a pressure of 756 Torr and a temperature of 15 °C. If the sample is compressed so that it has a volume of 11.2 L at 20 °C, what is the new pressure of the sample?

6. A sample of hydrogen gas occupies 27.5 L at 5 °C and 0.998 atm. How many moles of H_2 are in the sample?

7. A sample of oxygen gas is saturated with water vapor so that it is a mixture of oxygen and water vapor. If the total pressure of the sample is 745 Torr, what is the partial pressure of oxygen in the sample. A table of the vapor pressures of water can be found in Appendix 1.

8. What volume does 1.00 mole of a gas occupy at 25 °C and 1.00 atm?

9. The density of a gas at STP is 1.63 g/L. What is the molar mass or number of grams per mole of this gas?

10. What is the density of methane, CH_4, in grams per liter at 25 °C and 1.09 atm?

19 Solubilities and Solutions

Objective

The purposes of this experiment are to investigate the nature of solutions, consider some factors that affect solubility, and to prepare a solution of known molarity.

Discussion

Solutions are homogeneous mixtures in which the component particles are intermixed on the molecular or ionic level. When solid sugar is dissolved in water a liquid solution results. Water is the solvent. When a solution is prepared by mixing two chemicals of different physical states, the chemical that is of the same state as the resulting solution is called the solvent and the chemical that has been dissolved in the solvent is called the solute. If the two chemicals are of the same state, the solvent is usually considered to be the component present in the greatest amount. Generally, the solvent is the dissolver and the solute is the dissolved chemical.

Ammonia gas is very soluble in water. Some solutes, like ammonia, are quite soluble in water, and it is possible to prepare concentrated solutions of these solutes. Oxygen gas is somewhat soluble in water. Some solutes, like oxygen, are only slightly soluble in water, and some are so slightly soluble they are said to be insoluble. Many solid minerals are insoluble in water. Generally, the solubilities of chemicals in water range from soluble to slightly soluble to insoluble. The solubility of a chemical in water is often expressed in terms of the grams of solute that will dissolve in 100 g of water at a specified temperature.

A saturated solution contains enough solute so that a state of dynamic equilibrium exists between the dissolved and undissolved solute. A solution that contains less solute than a saturated solution under given conditions is called an unsaturated solution. Such a solution will dissolve more solute if more is added. A supersaturated solution contains more solute than a saturated solution. Such a solution is unstable and the addition of a small crystal, or often merely mixing the liquid, will cause the excess solute to crystallize, leaving a saturated solution.

The composition of a solution is often expressed in terms of a concentration that expresses the amount of solute dissolved in a specific amount of solution. A common concentration term is molarity, M, defined as the number of moles of solute per liter of solution.

$$\text{molarity} = \frac{\text{number of moles solute}}{\text{liter of solution}}$$

The molarity of a solution of known composition can be calculated by dividing the number of moles of solute in the solution by the volume of the solution in liters. Another concentration term is "percent by mass" which expresses the number of grams of solute per 100 g of solution. The percent by mass solute of a solution of known composition can be calculated by dividing the number of grams of solute by the number of grams of solution and multiplying by 100.

Laboratory Procedure

Space is provided so that you can record your observations. The observations should be noted on the report sheet along with the answers to any questions.

1. SUPERSATURATED SOLUTION

Place 2 g of sodium thiosulfate pentahydrate, $Na_2S_2O_3 \cdot 5H_2O$, in a test tube, add five drops of distilled water. Heat without boiling until the solid is completely dissolved. Make sure it all dissolves, otherwise the exercise will fail. Place the test tube in the test tube rack and let it cool for about an hour. (Go on with other parts of the experiment and return to this part later.) After cooling, gently shake the test tube and observe the solution. If nothing happens, drop a small crystal of $Na_2S_2O_3 \cdot 5H_2O$ into the tube and observe. Record your observations below. (A)

2. RELATIVE SOLUBILITIES

The purpose of this part of the experiment is to observe the differences in solubilities of three typical compounds. Obtain a 4 in. square of aluminum foil and place it on a wire gauze supported by a ring stand. Carry the foil on the gauze to the desk where dropper bottles of saturated solutions of silver chloride, calcium sulfate and sodium chloride are located. Carefully deposit one or two drops of each solution on the foil. Separate the drops of solution so that they do not run together and keep in mind which drops are which solution. Replace the gauze on the ring and carefully heat the foil using a low Bunsen flame until all of the solutions have evaporated. Describe your observations below and indicate how you would classify each of the solutes (soluble, slightly soluble, or insoluble). (B)

Solubilities of chemicals can be found in reference books. Fill in the following table using the <u>Handbook of Chemistry and Physics</u>. (C)

Solubility of	Cold Water	Temperature	Hot Water	Temperature
$AgCl$				
$CaSO_4$				
$NaCl$				

3. RECRYSTALLIZATION

This part of the experiment is intended to illustrate how a chemical can be purified by dissolving it in a solvent and then allowing the chemical to recrystallize from the solution. Place 5 mL of distilled water in a large test tube and add 3 g of ammonium chloride, NH_4Cl. Heat the solution until all of the solid has dissolved but do not let the solution boil. After the solid has dissolved, cool the solution by holding the test tube in the water tap. Describe your observations.(D)

How could the process of recrystallization be carried out to remove an insoluble impurity? Hint: Solids can be removed by filtration. (E)

Why is a small amount of soluble impurity usually removed when the recrystallized solid is filtered out of the liquid? (F)

4. SOLUBILITY OF GASES IN WATER

Water in contact with the atmosphere will dissolve slight amounts of oxygen and nitrogen gas from the air. Fill a 250-mL Erlenmeyer flask with tap water and insert a one-hold stopper. Make sure that the flask is full and that no air bubbles are trapped in the flask. Place an iron ring over the neck of the flask, hold a finger over the hole in the stopper, and invert the flask in a beaker of water. Support the flask with a ring so that the mouth is below the surface of the water in the flask. The apparatus is shown in Fig. 19-1. Carefully heat the water in the flask to near boiling, but do not allow it to boil. To avoid boiling, do not heat the flask at one point but move the flame around to heat the water evenly. As soon as you observe the release of the dissolved gases, stop the heating. Water does not decompose by this heating. Heating causes the dissolved oxygen and nitrogen gas to become less soluble. Record your observations. (G)

What do you conclude about the solubilities of these gases in water when the temperature rises? (H)

What chemicals are present in the bubbles? (I) Are they present in the same proportions as in the atmosphere? Why? (J)

Fill in the following table using the <u>Handbook of Chemistry and Physics</u>. (K)

Solubility of	Cold Water	Temperature	Hot Water	Temperature
Oxygen, O_2				
Nitrogen, N_2				

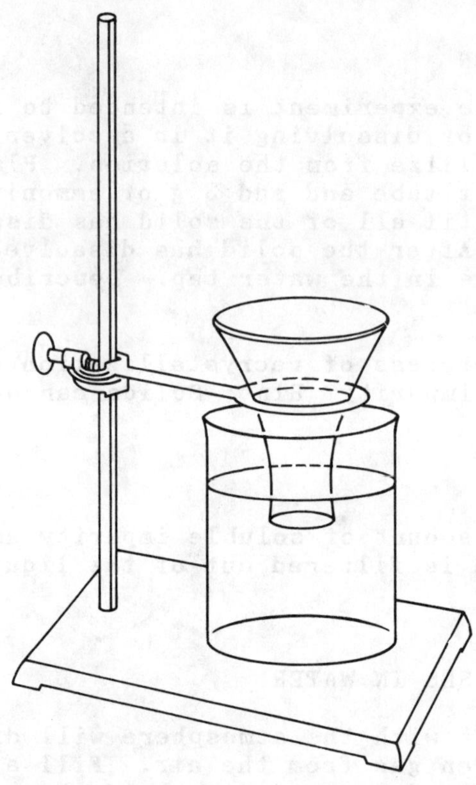

Fig. 19-1

5. THE CONCENTRATION OF A SOLUTION

The first step of this exercise involves the calibration of the volume of a flask. You will need a clean, dry 125-mL Erlenmeyer flask, a rubber stopper for the flask, and a clean, dry 200-mL beaker. Carefully attach a gummed label or a piece of tape around the neck of the flask so that the top edge of the label forms a horizontal line slightly above the place where the neck meets the body of the flask. Stopper the flask and weigh it to the nearest 0.01 g. (M) Obtain about 150-mL of distilled water in your 200-mL beaker. Pour the distilled water into the 125-mL flask and carefully fill it until the water level coincides with the top edge of the label marker. Make sure that the flask is dry on the outside. Stopper the flask and weigh it to the nearest 0.01 g. (L) Pour the distilled water in the flask back into the beaker.

To find the mass of the water in the flask, subtract the mass of the empty flask from the mass of the flask filled with water. (N) Using the density of 1.00 g/mL as the density of water, calculate the volume of water in the flask to three digits. (O) Convert this volume from milliliters to liters to give the calibrated volume of the flask. (P)

Obtain a sample of sodium chloride, NaCl, from your instructor or use a about a 5 gram sample from the bottle provided. Weigh a piece of weighing paper to the nearest 0.01 g. (R) Using the paper carefully, weigh the entire sodium chloride sample to the nearest 0.01 g. (Q) Subtract the mass of the paper to find the mass of the sodium chloride sample. (S) Be careful not to spill any of the weighed sample.

Carefully pour the sodium chloride sample into the calibrated 125-mL flask and take care not to spill any of the sample. Using the distilled water in the beaker, pour about 50 mL of distilled water into the flask, stopper it and gently swirl to dissolve the sodium chloride. Now, carefully add distilled water until the solution level coincides with the top edge of the label marker. Tightly stopper the flask and gently shake the solution to make it homogeneous. Make sure that the flask is not wet on the outside. Weigh the flask and the solution to the nearest 0.01 g. (T) Subtract the mass of the empty flask from the mass of the flask and the solution to find the mass of the solution. (U)

Using the mass of the sodium chloride (S), the mass of the solution (U), and the volume of the solution which is the same as the calibrated volume of the flask (P), calculate the molarity of the sodium chloride solution (V), the percent sodium chloride by mass in the solution (W), and the density of the sodium chloride solution (X). In the space provided on the report sheet, devise a label which describes the molarity of your solution. (Y)

19

REPORT SHEET

Name ________________

Lab Section ________________

Due Date ________________

EXPERIMENT 19 (Page 1)

1. Supersaturated Solution (A)

2. Relative Solubilities (B)

(C) Solubility of	Cold Water	Temperature	Hot Water	Temperature
AgCl				
$CaSO_4$				
NaCl				

3. Recrystallization

(D)

(E)

(F)

4. Solubility of Gases in Water

(G)

(H)

(I)

(J)

(K) Solubility of Cold Water Temperature Hot Water Temperature

Solubility of	Cold Water	Temperature	Hot Water	Temperature
Oxygen, O_2				
Nitrogen, N_2				

5. The Concentration of a Solution

Mass of flask and water ____________ (L)

Mass of flask ____________ (M)

Mass of water in flask ____________ (N)

Volume of water in flask ____________ (O)

Volume in liters ____________ (P)

Mass of NaCl and paper ____________ (Q)

Mass of paper ____________ (R)

Mass of NaCl sample ____________ (S)

EXPERIMENT 19

Name ________________

Lab Section ________________

Due Date ________________

EXPERIMENT 19 (Page 2)

Mass of flask and solution ____________ (T)

Mass of flask ____________ (M)

Mass of solution ____________ (U)

Data:

Mass of NaCl sample ____________ (S)

Mass of solution ____________ (U)

Volume of solution ____________ (P)

Calculations (show setups and answers)

(V) Molarity of NaCl solution

(W) Percent NaCl by mass

(X) Density of NaCl solution

(Y) Solution Label

19

QUESTIONS

Name ______________________________

Lab Section ______________________

Due Date _________________________

EXPERIMENT 19

1. Most solutions we work with in the laboratory are (circle one)

(a) unsaturated (b) saturated (c) supersaturated

2. The component of a solution sometimes designated as the dissolver is the (circle one)

(a) solution (b) solvent (c) solute

3. Some chemicals can be purified by removing impurities by (circle one)

(a) dissolving (b) recrystallization (c) heating

4. The solubility of air in water is decreased by (circle one)

(a) heating (b) cooling (c) mixing

5. A solution is prepared by dissolving 6.92 g of sodium sulfate, Na_2SO_4, in water to form 150 mL of solution. Calculate the molarity of the solution.

6. A water solution of glucose contains 32 g of glucose in 137 g of solution. Calculate the percent by mass of glucose in the solution.

20 Conductivities of Solutions

Objective

The purposes of this experiment are (1) to observe the conductivities of some solutions of chemicals and classify them as nonelectrolytes or strong or weak electrolytes, (2) to observe the changes in conductivity that take place when reactions involving ions occur, and (3) to observe the electrolysis of water.

Discussion

Solutions containing ions can conduct electricity. If two metal wires are connected to the terminals of a battery (or generator) one wire can be considered to be negatively charged and the other positively charged. If these wires are dipped into a solution containing ions, the positive ions (cations) are attracted to one wire and the negative ions (anions) are attracted to the other wire. The wires are called electrodes. The electrode that attracts the cations is called the cathode and the electrode that attracts the anions is called the anode. The external battery can serve as an electron pump and, if the battery supplies a sufficient driving force, chemical reactions involving the loss and gain of electrons occur at the electrodes. Electrons are gained by some species at the cathode and lost to form some other species at the anode. The migration of ions and the loss or gain of electrons provides a path for the flow of electrical current. Thus, the solution of ions is said to be conducting electricity. When a solution conducts electricity in this manner, a chemical is formed at the cathode and another at the anode. Such a process is called electrolysis.

By observing whether or not an aqueous solution of a chemical conducts, it is possible to see if ions are present in the solution. A chemical that forms an aqueous solution that conducts electricity is called an electrolyte. A chemical that forms an aqueous solution that does not conduct is called a nonelectrolyte. Many molecular chemicals are nonelectrolytes. A nonelectrolyte does not form ions when it dissolves. Electrolytes form ions when they dissolve. Solutions of some electrolytes are strong conductors while solutions of others are weak conductors. The difference is due to the extent to which chemicals form ions when they are dissolved in water. Soluble ionic chemicals are strong electrolytes. In dissolving they break up into the constituent anion and cation. For example, when $CaCl_2$ dissolves in water it separates into ions:

$$CaCl_2(s) \longrightarrow Ca^{2+} + 2Cl^-$$

In general the dissolving of an ionic compound can be represented as:

$$C_nA_m(s) \longrightarrow nC^{m+}(aq) + mA^{n-}(aq)$$

Some molecular compounds react completely with water when dissolved to form hydronium ion and an anion. Nitric acid is a strong electrolyte that reacts with water as shown by the following equation.

$$HNO_3 + H_2O \longrightarrow H_3O^+(aq) + NO_3^-(aq)$$

Some molecular chemicals react with water to a slight extent when dissolved to form some ions, but the major portion of the chemical remains in solution in the molecular form. These chemicals are weak electrolytes. Still other molecular chemicals dissolve in water and form no ions at all. These are the nonelectrolytes.

To indicate the nature of a solution of a chemical, it is best to give the formula of the major species present in the solution. Strong electrolytes are represented by the formulas of the ions they form in solution. Weak electrolytes and nonelectrolytes are represented by the molecular formula of the dissolved chemical. For example, a solution of calcium chloride contains Ca^{2+} and Cl^-, a solution of the weak electrolyte ammonia is represented as NH_3, and a solution of the nonelectrolyte hydrogen peroxide contains H_2O_2.

Laboratory Procedure

1. CONDUCTIVITY

Test the conductivities of the solutions and chemicals shown in the table in the report sheet and indicate whether they are good, weak or nonconductors. (A) A typical conductivity apparatus is pictured in Fig. 20-1. After testing the conductivities, give the formulas of the major species in the various solutions. Note that hydrochloric acid (HCl) in water and sulfuric acid (H_2SO_4) in water are like nitric acid mentioned in the discussion.

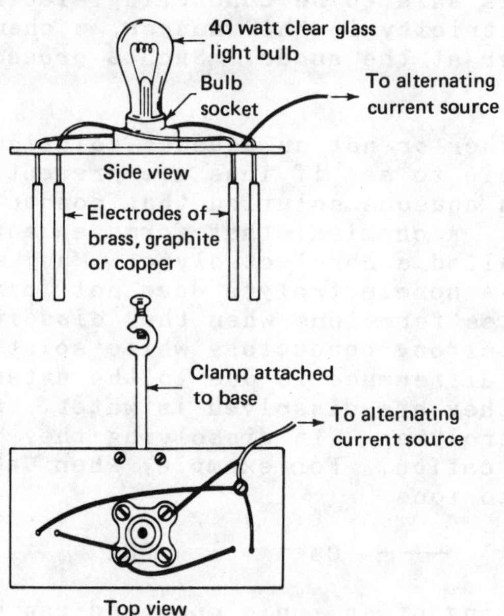

Fig. 20-1 Conductivity apparatus.

2. IONIC REACTIONS

Observe the conductivities of each of the following solutions. Then observe any change in conductivity that occurs when the solutions are mixed.

(a) Hydrochloric acid 0.1 M _______________ Sodium hydroxide 0.1 M _______________ (B) Change in conductivity as mixing occurs. (C)

The equation for the chemical reaction is:

$H_3O^+(aq) + Cl^-(aq) + Na^+(aq) + OH^-(aq) \longrightarrow Na^+(aq) + Cl^-(aq) + 2H_2O$

Explain why any change in conductivity was observed. (D)

(b) Acetic acid 0.1 M ___________ Aqua ammonia 0.1 M __________ (E) Change in conductivity as mixing occurs. (F)

The equation for the chemical reaction is:

$HC_2H_3O_2(aq) + NH_3(aq) \longrightarrow NH_4^+(aq) + C_2H_3O_2^-(aq)$

Explain why any change in conductivity was observed. (G)

(c) Copper(II) sulfate 0.1 M __________ Barium hydroxide 0.1 M __________ (H) Change in conductivity as mixing occurs. (I)

The equation for the chemical reaction is:

$Cu^{2+}(aq) + SO_4^{2-}(aq) + Ba^{2+}(aq) + 2OH^-(aq) \longrightarrow BaSO_4(s) + Cu(OH)_2(s)$

Explain why any change in conductivity was observed. (J)

(d) $CaCO_3$ (marble) chips in water __________ acetic acid 6 M __________ (K) Change in conductivity as mixing occurs. (L)

The equation for the chemical reaction is:

$CaCO_3(s) + 2HC_2H_3O_2(aq) \longrightarrow Ca^{2+}(aq) + 2C_2H_3O_2^-(aq) + CO_2(g) + H_2O$

Explain why any change in conductivity was observed. (M)

3. ELECTROLYSIS OF WATER (Optional)

Water can be decomposed into hydrogen and oxygen by a process of electrolysis using a special apparatus in which electricity is passed through a water sample. (See Fig. 20-2.) Give the balanced equation for the decomposition of water. (N)

Observe the electrolysis of water and describe your observations. (O)

What relative proportions of hydrogen and oxygen are produced during electrolysis? (P)

Explain how you can deduce the empirical formula of water based on your observations of the electrolysis of water. (Q)

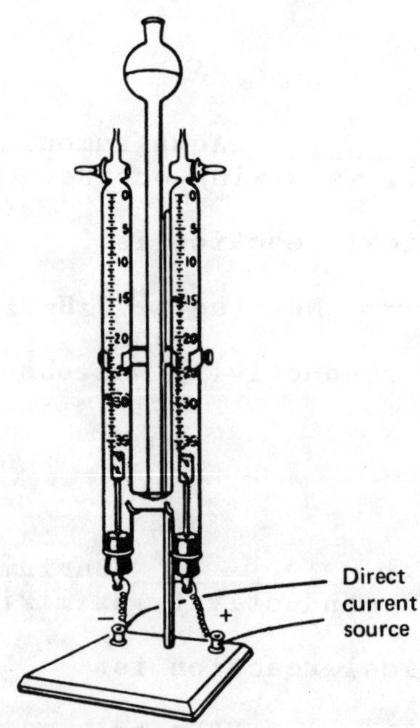

Fig. 20-2 Electrolysis of water apparatus.

20

REPORT SHEET

Name _______________

Lab Section _______________

Due Date _______________

EXPERIMENT 20

1. Conductivity

(A)

Solution or Chemical	Conductivity	Formula or Formulas of Major Species
Pure (distilled) water		
Tap water		
Acetic acid, 18 M		
Acetic acid, 0.1 M		
Aqua ammonia, 0.1 M		
Ethyl alcohol		
Ethyl alcohol solution		
Hydrochloric acid, 12 M		
Hydrochloric acid, 0.1 M		
Hydrogen sulfide solution		
Potassium sulfate, 0.1 M		
Sodium acetate, 0.1 M		
Sodium chloride		
Sodium chloride, 0.1 M		
Sodium hydroxide, 0.1 M		
Sulfuric acid, conc.		
Sulfuric acid, 0.1 M		
Sucrose		
Sucrose solution		

2. Ionic Reactions

(B) a. Hydrochloric acid 0.1 M ________Sodium hydroxide 0.1 M________

(C)

(D)

(E) b. Acetic acid 0.1 M ________ Aqua ammonia 0.1 M ________

(F)

(G)

(H) c. Copper(II) sulfate 0.1 M ______Barium hydroxide 0.1 M_________

(I)

(J)

(K) d. $CaCO_3$ (marble) chips in water _______Acetic acid 6M__________

(L)

(M)

3. Electrolysis of water

(N)

(O)

(P)

(Q)

20

QUESTIONS

Name ______________________________

Lab Section ______________________

Due Date _________________________

EXPERIMENT 20

1. Define the following terms:

   (a) Electrolyte

   (b) Weak Electrolyte

2. Predict the formula or formulas of the major species contained in solutions of the following compounds given the conductivity behavior.

(a) A solution of $MgBr_2$ a strong conductor.

(b) A solution of hydrochloric acid, HCl, is a strong conductor.

(c) A solution of NH_4NO_3 is a strong conductor.

(d) A solution of sucrose, $C_{12}H_{22}O_{11}$, is a nonconductor.

(e) A solution of phosphoric acid, H_3PO_4, is a weak conductor.

(f) A solution of ethyl alcohol, C_2H_6O, is a nonconductor.

20

QUESTIONS

Name ______________________________

Lab Section ______________________________

Due Date ______________________________

EXPERIMENT 20

1. Define the following terms:

(a) Electrolyte

(b) Weak Electrolyte

2. Predict the formula or formulas of the chief species contained in the solutions of the following compounds given the conductivity behavior.

(a) A solution of NaF is a strong conductor.

(b) A solution of hydrochloric acid, HCl, is a strong conductor.

(c) A solution of NH_4NO_3 is a strong conductor.

(d) A solution of sucrose, $C_{12}H_{22}O_{11}$, is a nonconductor.

(e) A solution of phosphoric acid, H_3PO_4, is a weak conductor.

(f) A solution of ethyl alcohol, C_2H_5OH, is a nonconductor.

21 Equilibrium and Le Châtelier's Principle

Objective

The purposes of this exercise are to experimentally observe some equilibrium systems and test how changes can affect the equilibrium reactions.

Discussion

A reversible chemical reaction is a reaction in which the reactants can form the products and the products can form the reactants. In a reversible reaction, both the forward and reverse reactions occur and act in opposition. When a reaction of this type occurs, it soon reaches a state of dynamic equilibrium. Dynamic equilibrium is established when the rate of the forward reaction equals the rate of the reverse reaction. At equilibrium, the reactions occur continually but a balance of rates exists. Consequently, when a reversible reaction is at equilibrium, specific concentrations of reactants and products are present. An equilibrium is denoted by a double arrow between reactants and products.

The concentrations of gaseous species and dissolved species involved in a reversible reaction become constant when the reaction reaches equilibrium. This is true even though the reactions are continually occurring. The balance in concentrations of gaseous or dissolved species can be expressed in terms of an equilibrium constant. The constant comes from the fact that at equilibrium the ratio of the concentrations of the products to the concentrations of the reactants must equal some fixed and constant value at a given temperature. The equilibrium constant expression for a reaction like

$$\begin{matrix} \text{reactants} & & \text{products} \\ 2A & \rightleftharpoons & B \end{matrix}$$

can be written as

$$K_{eq} = \frac{[B]}{[A]\,[A]} = \frac{[B]}{[A]^2}$$

where A and B are some gaseous or dissolved species, K_{eq} is the equilibrium constant and the square brackets [ ] represent concentrations in moles per liter. Note that, since species A is essentially two reactants, its concentration ends up to be squared in the expression. In general, the power used for any species in the equilibrium constant expression is its coefficient in the equation.

A reaction system in equilibrium will remain in equilibrium indefinitely unless something upsets the equilibrium. At equilibrium, a balance of concentrations exists and, if a concentration is changed in any way, the equilibrium can be upset. Le Châtelier's principle is a useful guide to predict what may happen when an equilibrium is upset. A statement of the principle is: When an equilibrium system is upset by a change in any factor affecting the equilibrium, the equilibrium shifts in a direction that tends to counteract the change.

At equilibrium, a balance of concentration exists. If any one of the concentrations is increased by adding an additional amount of a species, the equilibrium will shift in a direction which tends to decrease the concentration to keep the equilibrium balance. This is what the principle indicates. The idea is that the equilibrium shifts in one direction or the other to maintain the concentration balance. However, it is important to note that,in an equilibrium system the includes a solid or a liquid, that the concentration of the solid or liquid cannot be changed by adding more of the solid or liquid. On the other hand, the concentration of a gas or a dissolved species in an equilibrium system can be changed by adding more of the gas or dissolved species.

An equilibrium system not only has a balance of concentrations of species, but also has an energy balance. Energy or heat is another factor that can affect an equilibrium system. In a sense, energy can be viewed as a reactant or product. If the energy supply is changed by heating or cooling, the equilibrium according to Le Châtelier's principle will shift. If an equilibrium system is heated, the equilibrium shifts in a direction that tends to use up the energy. If the system is cooled, the equilibrium shifts in a direction that tends to provide energy.

Laboratory Procedure

1. EQUILIBRIUM AND LE CHÂTELIER'S PRINCIPLE

Several equilibrium systems will be investigated and the effect of a temperature or concentration change will be observed. Space is provided for your notes and observations. Your observations should be summarized on the report sheet.

(a) Saturated Sodium Chloride Solution.

Fill a large test tube with solid NaCl to a depth of about 3 cm. Add 10 mL of distilled water and gently heat the solution to boiling in a Bunsen flame. Stir the mixture with a glass rod. Set the solution aside to cool, go on with part (b), and return to this part later. Pour about 2 mL of the cooled solution into each of two small test tubes, and try not to transfer any solid salt. To the first tube add two or three drops of concentrated hydrochloric acid ($H_3O^+ + Cl^-$). To the second tube add a large crystal of rock salt. Describe any changes that occur in either tube. (A)

The equilibrium in a saturated solution of sodium chloride is:

$$NaCl(s) \rightleftharpoons Na^+(aq) + Cl^-(aq)$$

Using this equilibrium and Le Châtelier's principle, explain your observations.

(b) Nitrogen Dioxide - Dinitrogen Tetroxide Equilibrium.

Nitrogen dioxide, NO_2, is a red-brown gas and dinitrogen tetroxide, N_2O_4, is a colorless gas. The two gases exist in equilibrium. Put some ice with water in a 400-mL beaker. Obtain a sealed tube of the equilibrium mixture from the stockroom and note the color of the mixture. Place the tube in the ice-water mixture for a few minutes. Go on to part (c) and return to this part later. Note any color change in the tube after it has cooled. Remove the tube from the ice water, dry it and, by gently rubbing with your hands, warm the tube. Note any color changes. Write an equilibrium equation for the reversible reaction involved in this section. Indicate which side of the equation requires energy by writing in the word "energy" on the proper side of the equation. Using Le Châtelier's principle, explain your observations. (B)

(c) Saturated Ammonium Chloride Solution.

Fill a large test tube with solid NH_4Cl to a depth of about 3 cm. Add 5 mL of distilled water and stir the mixture with a glass rod. After mixing, carefully heat the tube in the Bunsen flame until it just begins to boil. Keep heating without boiling until all of the solid dissolves. Hold the test tube under a cold water tap and allow the solution to cool. Note the tube as it cools or after cooling. Compare this to what happened upon heating. Write an equilibrium equation for a saturated solution of ammonium chloride. (This is similar to the equation in part (a).) Write the word "energy" on the proper side of the equation and explain your observations using Le Châtelier's principle. (C)

(d) Iron(III) Ion and Thiocyanate Ion in Equilibrium.

Fill two test tubes each with 10 mL of distilled water. Add 5 drops of 0.1 M $FeCl_3$ to one tube and 1 drop of 1 M NH_4CNS to the other. Pour one solution into the other and describe your results. The equilibrium involved is:

$$Fe^{3+}(aq) + CNS^{-}(aq) \rightleftharpoons FeCNS^{2+}(aq)$$

The $FeCNS^{2+}$ is red in color.

Pour one third of the solution into each of three test tubes. Keep one for color comparison. To one tube add several drops of 0.1 M $FeCl_3$ and describe any change. To another tube add a few drops of 1 M NH_4CNS and describe any change. Use the above equilibrium and Le Châtelier's principle to explain any changes. (D)

(e) Bromthymol Blue in Acid and Base Solutions.

Bromthymol blue is a complex organic compound which we will represent as HBB. In water it exists in the equilibrium

$$HBB(aq) + H_2O \rightleftharpoons H_3O^+(aq) + BB^-(aq)$$

The HBB has a yellow color and the BB^- ion has a blue color.

Add about 5 mL of distilled water to a large test tube followed by 6 drops of the bromthymol blue solution provided. Pour about 1 mL of a 6 M hydrochloric acid (H_3O^+ + Cl^-) solution into the tube. Describe any change. Now, pour about 2 mL of a 6 M NaOH solution into the tube. Describe any change. The hydroxide ion in the NaOH reacts with hydronium ion by the reaction:

$$H_3O^+(aq) + OH^-(aq) \longrightarrow 2H_2O$$

This reaction decreases the concentration of hydronium ion. Use the above equilibrium reaction and Le Châtelier's principle to explain your observations. (E)

(f) Chromate Ion - Dichromate Ion Equilibrium.

In acid solutions, chromate ion and dichromate ion exist in equilibrium according to the equation shown below.

$$2CrO_4^{2-}(aq) + 2H_3O^+(aq) \rightleftharpoons Cr_2O_7^{2-}(aq) + 3H_2O$$

Record the colors of the K_2CrO_4 and $K_2Cr_2O_7$ solutions that are on display. Pour about 2 mL of 0.1 M K_2CrO_4 solution into each of two test tubes. To one test tube add 2 drops of 6 M nitric acid (H_3O^+ + NO_3^-) and note any changes. To the other test tube add 2 drops of 3 M sulfuric acid (H_3O^+ + HSO_4^-) and note any changes. To each tube add 6 M NaOH drop by drop until the solution changes to the original color. See part (e) for a note on the reaction of hydroxide ion and hydronium ion. Use the above equilibrium equation and Le Châtelier's principle to explain your observations. (F)

2. EQUILIBRIUM EXPRESSIONS

Equilibrium expressions show an equilibrium constant, K_{eq}, as the ratio of the concentrations of the reactants. For the equilibrium reactions in parts (b), (d), (e) and (f), write an appropriate equilibrium constant expression. Remember that only species in solution and gases are included in the expression. Solids and liquids are never included in the expression and water can be excluded from expressions involving equilibrium in water solutions. (G)

(b)

(d)

(e)

(f)

21

REPORT SHEET

Name _______________

Lab Section _______________

Due Date _______________

EXPERIMENT 21

1. Equilibrium and Le Châtelier's Principle

(a) Saturated Sodium Chloride Solution

equilibrium equation:

(A)

(b) Nitrogen Dioxide - Dinitrogen Tetroxide Equilibrium

equilibrium equation:

(B)

(c) Saturated Ammonium Chloride Solution

equilibrium equation:

(C)

(d) Iron(III) Ion and Thiocyanate Ion in Equilibrium

equilibrium equation:

(D)

(e) Bromthymol Blue in Acid and Base Solutions

equilibrium equation:

(E)

(f) **Chromate - Dichromate Ion Equilibrium**

equilibrium equation:

(F)

2. **Equilibrium Expressions**

(G)

(b)

(d)

(e)

(f)

21

QUESTIONS

Name ______________________________

Lab Section ____________________

Due Date _______________________

EXPERIMENT 21

1. Given the following chemical equilibrium,

$$CO_2(g) + H_2(g) \rightleftharpoons CO(g) + H_2O(g)$$

use Le Châtelier's principle to explain the following observations:

(a) Upon heating, the equilibrium shifts towards the CO/H_2O side.

(b) A decrease in the concentration of H_2 causes the equilibrium to shift towards the CO_2/H_2 side.

2. Phenol red is a complex organic compound sometimes used as an indicator for testing swimming pool water. We can represent phenol red by the symbolic formula HPR. In water phenol red is involved in the following equilibrium reaction:

$$HPR(aq) + H_2O \rightleftharpoons H_3O^+(aq) + PR^-(aq)$$

HPR has a yellow color and PR^- has a red color.

(a) If a sample of water with a relatively high concentration of H_3O^+ is tested with phenol red, what color will result?

(b) If a sample of water with a relatively low concentration of H_3O^+ is tested with phenol red what color will result?

3. Write equilibrium constant expressions for

(a) The equilibrium reaction in question 1.

(b) The equilibrium reaction in question 2.

QUESTIONS

Name ____________________

Lab Section ____________________

Due Date ____________________

ASSIGNMENT

1. [illegible] the following chemical equilibrium:

[illegible] $\rightleftharpoons$ [illegible] (g) + H_2O [illegible]

Use Le Chatelier's principle to explain [illegible] following observations:

(a) [illegible] the equilibrium to shift towards the [illegible]

(b) A decrease in the [illegible] cause the equilibrium to shift towards the [illegible] side.

2. [illegible] compound [illegible] indicator for testing [illegible] water. We can represent [illegible] by the symbolic formula HPh. In [illegible] the following equilibrium reaction:

HPh(aq) + H_2O $\rightleftharpoons$ H_3O^+(aq) + Ph^-(aq)

HPh has a yellow [illegible] color and [illegible] red [illegible].

(a) If a sample of water with relatively high concentration of [illegible] is tested with phenolphthalein, what color will result?

(b) If a sample of water with [illegible] low concentration of [illegible] is tested with phenolphthalein, what color will result?

3. [illegible] for the following [illegible] (a) [illegible]

(a) The equilibrium reaction in Question 1

(b) The equilibrium reaction in Question 2

22 Precipitation Reactions

Objective

The purpose of this experiment is to carry out some precipitation reactions, observe them and to write the net-ionic equations for the reactions.

Discussion

Many substances dissolve in water to form solutions. Most aqueous solutions are quite stable and serve as convenient sources of dissolved substances. In fact, most solutions will last indefinitely if stored in a sealed container. Solutions may contain ions (cations and anions), molecular species, or mixtures of ions and molecular species. When two solutions are mixed, the species contained in the solutions intermingle and may react chemically.

Certain ionic substances are insoluble in water and, thus, it is not possible to prepare solutions of these substances. On the other hand, many substances are quite soluble. The solubilities of common ionic substances are summarized in Table 22-1. Usually, when soluble ionic substances dissolve in water, the ions dissociate and enter the solutions as ions (e.g., $NaCl(s) \longrightarrow Na^+(aq) + Cl^-(aq)$). When solutions containing ions are mixed, the ions intermingle and some may react to form an insoluble solid called a precipitate. That is, if any pair of ions in the mixture are the constituents of an insoluble solid, these ions will combine to form the solid. Once formed, the solid will usually settle to the bottom of the solution or precipitate out of solution. As an example, suppose a solution of magnesium chloride is mixed with a solution of sodium hydroxide. The species that are mixed are the separated ions of the ionic compounds:

$$Mg^{2+} + Cl^- \qquad \text{and} \qquad Na^+ + OH^-$$

The magnesium ions and hydroxide ions react to form insoluble magnesium hydroxide (See Table 22-1.) as represented by the equation:

$$Mg^{2+} + 2OH^- \longrightarrow 2Mg(OH)_2$$

The sodium ion and the chloride ion do not react and, thus, are not included in the equation for the reaction. They are called spectator ions. An equation showing the ions and other species that react and are formed is called a net-ionic equation. A net-ionic equation must balanced both chemically (the same number of atoms of each element on each side) and electrically (the same amount of positive and negative charge on each side of the equation).

To write an equation for a precipitation reaction, first write down the species present in the solutions that are mixed. Consult

Table 22-1 to see if any of the possible combinations of ions will form an insoluble compound. Decide which ions react to form a precipitate, write these down along with the product and balance the equation. The ions that do not react are not included in the equation.

Table 22-1

Solubility List

Anion	*Cations That Form Precipitates*	*Cations That Do Not Form Precipitates*
NO_3^-	None	All other common
$C_2H_3O_2^-$	None	All other common
SO_4^{2-}	Ca^{2+}, Sr^{2+}, Ba^{2+}, Pb^{2+}	Most other common
Cl^-, Br^-, I^-	Ag^+ , Hg_2^{2+}, Pb^{2+}	Most other common
OH^-	Most	Na^+, K^+, Ba^{2+}
F^-	Mg^{2+}, Ca^{2+}, Sr^{2+}, Ba^{2+}, Pb^{2+}	Most other common
CO_3^{2-}, CrO_4^{2-}, PO_4^{3-}	Most	Na^+, K^+, NH_4^+
Almost all ionic compounds of Na^+, K^+ and NH_4^+ are soluble		

Laboratory Procedure

1. SOME PRECIPITATION REACTIONS

For each of the parts, mix the solutions indicated, record any evidence of a reaction, briefly describe the color and appearance of any precipitate, and write a balanced net-ionic equation for any reaction. Space is provided for your observations and equations. These should also be recorded on the Report Sheet. Carry out the mixing in small, clean test tubes and use a clean glass rod for stirring the mixtures.

(a) Place about 1 mL of a 0.1 M sodium chloride solution into a test tube. Pour in about 1 mL of a 0.1 M silver nitrate solution. Hint: The ions that are mixed are $Na^+ + Cl^-$ and $Ag^+ + NO_3^-$.
(A)

(b) Place about 1 mL of a 0.1 M barium nitrate solution into a test tube. Pour in about 1 mL of a 0.1 M sodium sulfate solution.
(B)

(c) Place about 1 mL of a 0.1 M silver nitrate solution into a test tube. Pour in about 1 mL of a 0.1 M potassium iodide solution.
(C)

(d) Place about 1 mL of a 0.1 M iron(III) chloride solution in a test tube. Pour in about 2 mL of a 0.1 M sodium hydroxide solution.
(D)

(e) Place about 1 mL of a 0.1 M sodium chloride solution into a test tube. Pour in about 1 mL of a 0.1 M potassium nitrate solution.
(E)

(f) Place about 1 mL of a 0.1 M sodium carbonate solution into a test tube. Pour in about 1 mL of a 0.1 M calcium chloride solution.
(F)

(g) Place about 1 mL of a 0.1 M barium hydroxide solution into a test tube. Pour in about 1 mL of a 0.1 M copper(II) sulfate solution.
(G)

(h) Place about 1 mL of a 0.1 M barium hydroxide solution into a test tube. Pour in about 1 mL of a 0.1 M sodium phosphate solution.
(H)

2. FOUR UNLABELED BOTTLES (OPTIONAL)

Four bottles of solutions are provided without labels. They are marked A,B, C and D for reference. The possible contents of the bottles are solutions of $AgNO_3$, NaCl, KI, and $NaNO_3$. Each bottle is one of these solutions. By mixing various combinations of the four solutions, they can be identified by observing any precipitation reactions that occur. Using the solubility list and your knowledge of the appearance of certain precipitates gained in part one of this exercise, you can deduce which bottle contains which solution.

To begin, consider the prediction of any reactions that will occur when any two of the above solutions are mixed. Give the balanced net-ionic equations for any reactions. Not all combinations will give reactions. The six possible mixtures are:

(a) $AgNO_3$ solution and NaCl solution
(I)

(b) $AgNO_3$ solution and KI solution
(J)

(c) $AgNO_3$ solution and $NaNO_3$ solution
(K)

(d) NaCl solution and KI solution
(L)

(e) NaCl solution and $NaNO_3$ solution
(M)

(f) KI solution and $NaNO_3$ solution
(N)

To determine the contents of the bottles, use the following approach. Place six small, clean test tubes in a test tube rack. Label the tubes or the rack so that each tube can be distinguished as either AB, AC, AD, BC, BD or CD. Take the test tubes to the set of unlabeled bottles and add the following:

Tube AB	10 drops of solution A
Tube AC	10 drops of solution A
Tube AD	10 drops of solution A
Tube AB	10 drops of solution B
Tube BC	10 drops of solution B
Tube BD	10 drops of solution B
Tube AC	10 drops of solution C
Tube BC	10 drops of solution C
Tube CD	10 drops of solution C
Tube AD	10 drops of solution D
Tube BD	10 drops of solution D
Tube CD	10 drops of solution D

Make sure that the contents of each tube are well mixed and describe the appearance and color of any precipitates that form. Using your results, identify the contents of the four bottles. (O)

A ________________ B ________________

C ________________ D ________________

(P) Explain the reasons for your choice.

22

REPORT SHEET

Name ________________

Lab Section ________________

Due Date ________________

EXPERIMENT 22

1. Some Precipitation Reactions (observations and equations)

(A) a.

(B) b.

(C) c.

(D) d.

(E) e.

(F) f.

(G) g.

(H) h.

2. Four Unlabeled Bottles

(I) (a) $AgNO_3$ solution and NaCl solution

(J) (b) $AgNO_3$ solution and KI solution

(K) (c) $AgNO_3$ solution and $NaNO_3$ solution

(L) (d) NaCl solution and KI solution

(M) (e) NaCl solution and $NaNO_3$ solution

(N) (f) KI solution and $NaNO_3$ solution

(O) A ____________________ B ____________________

C ____________________ D ____________________

(P)

22

QUESTIONS

Name ______________________________

Lab Section ______________________

Due Date _________________________

EXPERIMENT 22

Give balanced net-ionic equations for any reactions you would predict to occur when the following solutions are mixed. Use a solubility list.

(a) A KCl solution is added to a $Pb(NO_3)_2$ solution.

(b) A NH_4Cl solution is added to a KNO_3 solution.

(c) A KOH solution is added to a $MgCl_2$ solution.

(d) A $Pb(NO_3)_2$ solution is added to a NaBr solution.

(e) A $SrCl_2$ solution is added to a Na_2SO_4 solution.

(f) A KBr solution is added to a $AgNO_3$ solution.

(g) A NaOH solution is added to a KCl solution.

(h) A NaF solution is added to a $Ba(NO_3)_2$ solution.

(i) A $ZnSO_4$ solution is added to a $Ba(OH)_2$ solution.

QUESTIONS

Name ______________________

Lab Section ______________________

Due Date ______________________

EXPERIMENT 22

Give balanced net ionic equations for any reactions you would predict to occur when the following solutions are mixed. Use a solubility list.

(a) A HCl solution is added to a [illegible] solution.

(b) A NH_4Cl solution is added to a [illegible] solution.

(c) A KOH solution is added to a [illegible] solution.

(d) A $Pb(NO_3)_2$ solution is added to a [illegible] solution.

(e) A [illegible] solution is added to a [illegible] solution.

(f) A KBr solution is added to a [illegible] solution.

(g) A NaOH solution is added to a HCl solution.

(h) A [illegible] solution is added to a [illegible] solution.

(i) A [illegible] solution is added to a [illegible] solution.

23 Acids and Bases

Objective

The purpose of this experiment is to observe the properties and reactions of the aqueous solution of some common acids and bases.

Discussion

Acids and bases are very common chemicals and we use various acid solutions and base solutions in the laboratory. According to the Brønsted-Lowry theory of acids and bases, an acid is a species that has a tendency to lose a proton (H^+) in a chemical reaction. An acid is a proton donor. A base is a species that has a tendency to gain a proton in a chemical reaction. A base is a proton acceptor. These definitions can be illustrated by a general equation.

$$H\text{-}A + B \longrightarrow A^- + H\text{-}B^+$$

in which the acid, HA, loses a proton to the base, B. An actual example is represented by the equation

$$HCl(g) + H_2O \longrightarrow H_3O^+(aq) + Cl^-(aq)$$

in which the acid hydrogen chloride loses a proton to the base water to form hydronium ion and chloride ion. When an acid reacts with a base and a proton transfer occurs, the reaction is called an acid-base reaction. An acid loses a proton to a base which is capable of bonding to the proton more strongly than the original acid. Thus, we say that an acid, HA, can lose a proton to leave a weaker base, A^-, and a base, B, can gain a proton to form a weaker acid, HB^+. Each acid has a corresponding base and each base will have a corresponding acid. An acid and base that are related by the loss of a proton are called a conjugate acid-base pair. The acid-base pairs are labeled in the following equation.

$$HCl(g) + H_2O \longrightarrow H_3O^+(aq) + Cl^-(aq)$$

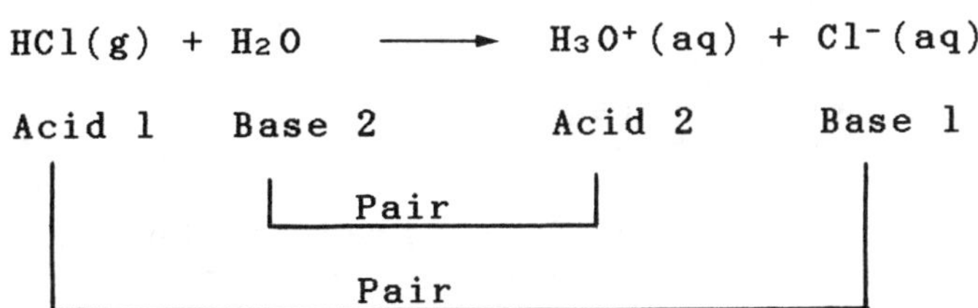

An acid-base reaction will always involve an acid losing a proton to a base to form the conjugate base of the acid and the conjugate acid of the base.

Many common chemicals act as acids or bases. Ammonia dissolved in water to form aqueous ammonia, $NH_3(aq)$, is a base. The soluble ionic compounds sodium hydroxide, NaOH, potassium hydroxide, KOH, and barium hydroxide, $Ba(OH)_2$, dissolve to form hydroxide ion, $OH^-(aq)$, and the corresponding metal ion. The hydroxide ion is a base. The hydrogen-containing compounds HCl, H_2SO_4, HNO_3, H_2S, $H_2C_2O_4$, $HC_2H_3O_2$, and H_3PO_4 are a few of the many acids that dissolve in water to form acidic solutions. Some acids are called strong acids since, when they dissolve in water, they react completely with water to form hydronium ion and the conjugate base of the acid. In general, a strong acid, H-X, will react in water as

$$H\text{-}X + H_2O \longrightarrow H_3O^+(aq) + X^-(aq)$$

Thus, the aqueous solutions of the strong acids will always contain hydronium ion and the conjugate base of the acid. Other acids only react slightly with water when they are mixed with water.

$$H\text{-}W + H_2O \rightleftharpoons H_3O^+(aq) + W^-(aq)$$

These acids are called weak acids. Solutions of weak acids will contain relatively low concentrations of hydronium ion and the conjugate base of the acid, and the major species in solution will be the acid. Thus, solutions of weak acids are represented by the formula of the acid. The species present in some common acid and base solutions are shown in Table 23-1.

When a solution of an acid is mixed with a solution of a base, a reaction will occur if a weaker acid and base can be formed. See Table 23-2 for relative strengths of acids and bases. To write a net-ionic equation representing an acid-base reaction, we write the acid and base that react on one side and the conjugate acid and base on the other. Table 23-2 can be used to predict the reaction. For example, consider the acid-base reaction that occurs when a solution of sodium hydroxide is mixed with a solution of the weak acid hydrogen fluoride, HF. Hydrogen fluoride solutions are toxic and dangerous. CAUTION: DO NOT WORK WITH HYDROGEN FLUORIDE SOLUTIONS IN THE LABORATORY. First, write the formulas of the species present in the solutions. The solution of the ionic compound sodium hydroxide contains

$$Na^+(aq) + OH^-(aq)$$

and the hydrogen fluoride solution contains HF(aq) since HF is a weak acid. (See Table 23-2.) When we mix these solutions, we are mixing the species

$$Na^+(aq) + OH^-(aq) + HF(aq)$$

Finding HF and OH^- in Table 23-2, we predict that they will react to give the weaker acid, H_2O, and the weaker base, F^-.

$$HF(aq) + OH^-(aq) \rightleftharpoons H_2O + F^-(aq)$$

As another example consider the acid base reaction that occurs when nitric acid and potassium hydroxide solutions are mixed. First, write down the species present in the solutions which are mixed. Use Table 23-1 as a guide. The major species present in nitric acid and potassium hydroxide are:

$$H_3O^+(aq) + NO_3^-(aq) + K^+(aq) + OH^-(aq)$$

In Table 23-2 we note that H_3O^+ is an acid and OH^- is a base. The reaction that occurs when a solution of a strong acid is added to a solution of a hydroxide is always

$$H_3O^+(aq) + OH^-(aq) \rightleftharpoons 2H_2O$$

Such a reaction forms water and is called a neutralization reaction. In general, an acid will react with any base that is below it in the acid-base table.

Indicators are chemicals that take on different colors depending upon the concentration of hydronium ion or hydroxide ion in an aqueous solution. Some indicators can be used to indicate whether a solution is acidic or basic. The indicator called litmus turns blue in a solution of a base and red in a solution of an acid. The indicator phenolphthalein is colorless in acidic solutions and turns pink in basic solutions. This indicator can be used to observe the completion of some acid-base reactions by adding indicator to an acid solution and then adding base solution until a pink color is noted. Of course, the appearance of the pink would indicate the change to a basic solution.

Table 23-1
Some Acid and Base Solutions

Strong Acids		
sulfuric acid	$H_3O^+ + HSO_4^-$	
nitric acid	$H_3O^+ + NO_3^-$	
hydrochloric acid	$H_3O^+ + Cl^-$	
Weak Acids		
phosphoric acid	$H_3PO_4(aq)$	
carbon dioxide	$CO_2(aq)$	(Use $CO_2 + H_2O$ for acid-base reactions)
acetic acid	$CH_3COOH(aq)$ or $HC_2H_3O_2(aq)$	
dihydrogen phosphate ion	$H_2PO_4^-(aq)$	(Prepared by dissolving NaH_2PO_4 or KH_2PO_4 in water.)
ammonium ion	$NH_4^+(aq)$	(Prepared by dissolving NH_4Cl in water.)
Bases		
benzoate ion	$C_6H_5COO^-(aq)$	(Prepared by dissolving a soluble compound like NaC_6H_5COO in water.)
ammonia	$NH_3(aq)$	
carbonate ion	$CO_3^{2-}(aq)$	(Prepared by dissolving a soluble compound like Na_2CO_3 in water.)
hydroxide ion	$OH^-(aq)$	(Prepared by dissolving NaOH or KOH in water.)

Table 23-2
Acid-Base Table

	Acids	Bases	
Sulfuric acid	H_2SO_4	HSO_4^-	Hydrogen sulfate ion
Hydrogen chloride	HCl	Cl^-	Chloride ion
Nitric acid	HNO_3	NO_3^-	Nitrate ion
Hydronium ion	H_3O^+	H_2O	Water
Oxalic acid	$H_2C_2O_4$	$HC_2O_4^-$	Hydrogen oxalate ion
Hydrogen sulfate ion	HSO_4^-	SO_4^{2-}	Sulfate ion
Phosphoric acid	H_3PO_4	$H_2PO_4^-$	Dihydrogen phosphate ion
Hydrogen fluoride	HF	F^-	Fluoride ion
Hydrogen oxalate ion	$HC_2O_4^-$	$C_2O_4^{2-}$	Oxalate ion
Acetic acid	$HC_2H_3O_2$	$C_2H_3O_2^-$	Acetate ion
Carbon dioxide (*aq*)	$(CO_2 + H_2O)$	HCO_3^-	Hydrogen carbonate ion
Hydrogen sulfide	H_2S	HS^-	Hydrogen sulfide ion
Dihydrogen phosphate ion	$H_2PO_4^-$	HPO_4^{2-}	Hydrogen phosphate ion
Hydrogen sulfite ion	HSO_3^-	SO_3^{2-}	Sulfite ion
Ammonium ion	NH_4^+	NH_3	Ammonia
Hydrogen cyanide	HCN	CN^-	Cyanide ion
Hydrogen carbonate ion	HCO_3^-	CO_3^{2-}	Carbonate ion
Hydrogen phosphate ion	HPO_4^{2-}	PO_4^{3-}	Phosphate ion
Hydrogen sulfide ion	HS^-	S^{2-}	Sulfide ion
Water	H_2O	OH^-	Hydroxide ion

Acids are listed according to decreasing strength. Bases are listed according to increasing strength.

Laboratory Procedure

1. ACIDIC AND BASIC SOLUTIONS

Using nine clean test tubes, obtain about 0.5 mL of each of the solutions listed in part 1 on the report sheet. Place the test tubes in your test tube rack. Obtain nine pieces of red litmus paper and nine pieces of blue litmus paper and place the paper on a towel. Use a glass stirring rod to transfer a drop of the hydrochloric acid solution to the end of one piece of red litmus and a drop to the end of one piece of blue litmus. Record your observations of the color change in the space provided on the report sheet. Wipe off the stirring rod and repeat using the other solutions. Be sure to wipe off the rod each time and to record the results.

To each of the various solutions, add a drop of phenolphthalein indicator. Record the color of the indicator in each solution in the space provided on the report sheet. (A)

2. ACID-BASE REACTIONS

For each of the following, give the formulas of the major species present in the solutions before mixing and give a balanced equation for any reaction. For your convenience, you do not have to include the parenthetical aq after each formula. Refer to Tables 23-1 and 23-2 as needed. For convenience, calibrate several small test tubes by pouring 2 mL of water in them and marking the 2 mL level with tape or waxed pencil. Rinse test tubes with distilled water after each use. For litmus tests, use strips of litmus paper on which drops of the solutions can be tested using a clean stirring rod. Space is provided for your observations and equations in each part. The observations and equations should also be recorded on the Report Sheet.

(a) Pour 2 mL of a 1 M hydrochloric acid solution into one clean test tube and slightly more than 2 mL of a 1 M sodium hydroxide solution in another. Put one drop of phenolphthalein indicator into the acid solution and then insert a thermometer into the solution. Note the temperature._________________

Pour the base solution into the acid solution and note any evidence for a reaction.

Temperature ______________ Color ________(B)

The formulas for the major species in the two solutions before mixing are $H_3O^+ + Cl^-$ and $Na^+ + OH^-$.

Write a balanced net-ionic equation for the acid-base reaction that took place. (C)

(b) Place 2 mL of a 1 M acetic acid solution in a clean test tube and test with litmus. Pour slightly more than 2 mL of a 1 M sodium hydroxide solution into the acetic acid. Cork, shake vigorously and test with litmus. (D)

Litmus tests:

The species mixed are: $HC_2H_3O_2$ and $Na^+ + OH^-$

Balanced equation:

(c) Place 2 mL of a 1 M sodium acetate solution in a test tube and test with litmus. Add slightly more than 2 mL of a 1 M hydrochloric acid solution to the tube. Cork, shake vigorously and test with litmus. (E)

Litmus tests:

The species mixed are: $Na^+ + C_2H_3O_2^-$ and $H_3O^+ + Cl^-$

Balanced equation:

(d) Place 2 mL of a 1 M ammonia solution in a test tube and test with litmus. Add slightly more than 2 mL of a 1 M hydrochloric acid solution to the tube. Cork, shake vigorously and test with litmus. (F)

Litmus tests:

The species mixed are: NH_3 and H_3O^+ + Cl^-

Balanced equation:

(e) Place 2 mL of a 1 M ammonium chloride solution in a test tube and test with litmus. Add slightly more than 2 mL of a 1 M sodium hydroxide solution to the tube. Cork, shake vigorously and test with litmus. (G) Avoid breathing the fumes.

Litmus tests:

The species mixed are: NH_4^+ + Cl^- and Na^+ + OH^-

Balanced equation:

(f) Place a small sample of solid benzoic acid about the size of a dried pea in a test tube. Add about 1 mL of distilled water, cork and shake vigorously. Comment on the solubility of the acid in water. Pour about 2 mL of a 1 M sodium hydroxide solution in the test tube, cork and shake vigorously for a few minutes. Describe what happens and test with litmus. (H)

Solubility in water:

Litmus test:

Give a balanced equation to show what happened. (Hint: benzoate ion is a product of the reaction. See Table 23-1.)

(g) Place 1 mL of a 1 M sodium benzoate (See Table 23-1) solution in a test tube and test with litmus. Add 2 mL of a 1 M hydrochloric acid solution to the tube, cork and shake. Describe the results. (I)

Litmus test:

Result:

Give a balanced equation to show what happened. (Hint: benzoic acid is insoluble in water.)

(h) Pour 2 mL of a 1 M sodium carbonate solution into a small beaker and test with litmus. Carefully and slowly add about 5 mL of a 1 M hydrochloric acid solution and describe the results. (Hint: Two acid-base reactions occur. The product of the first reaction is involved in the second reaction and the bubbles are CO_2 gas.) (J)

Litmus test:

Result:

The species mixed are:

Balanced equation:

(i) Pour 2 mL of a carbonated water (CO_2 in water) solution into a test tube. Gently shake and describe the solution. Add about 4 mL of a 1 M sodium hydroxide solution, cork and shake vigorously. Describe the solution. (K)

Description:

Give a balanced equation to show what happened.

(j) Place one aspirin tablet in a test tube. Aspirin is acetyl salicylic acid, $HC_9H_7O_4$, a solid weak acid. Add about 2 mL of distilled water to the tube and gently shake. Heat the tube in a Bunsen flame until boiling occurs. Now, cool the contents by running cold tap water over the outside of the tube. Comment on the solubility of acetyl salicylic acid in hot and cold water. Add about 4 mL of a 1 M sodium hydroxide solution, cork and shake. Describe what happens. (Hint: One product is acetyl salicylate ion, $C_9H_7O_4^-$. It is soluble.) (L)

Solubility:

Description:

Give a balanced equation to show what happened.

The second reaction and the bubbles are CO_2 (gas), [illegible] (l)→

Litmus test:

Result:

The solution mixed with:

Balanced equation:

(i) Pour 2 mL of a carbonated water (CO_2 in water solution) into a test tube. Gently shake and describe the solution. Add about 1 mL of a 1 M sodium hydroxide solution, cork and shake vigorously. Describe the solution. (i)

Description:

Give a balanced equation to show what happened.

(j) Place one aspirin tablet in a test tube. Aspirin is acetylsalicylic acid, $HC_9H_7O_4$, a solid weak acid. Add about 2 mL of distilled water to the tube and gently shake. Heat the tube in a Bunsen flame until boiling occurs. Now, cool the contents by running cold tap water over the outside of the tube. Comment on the solubility of acetylsalicylic acid in hot and cold water. Add about 3 mL of a 1 M sodium hydroxide solution, cork and shake. Describe what happens. (Hint: One product is acetylsalicylate ion, $C_9H_7O_4^-$, it is soluble.) (j)

Solubility:

Description:

Give a balanced equation to show what happened.

23

REPORT SHEET

Name ________________

Lab Section ________________

Due Date ________________

EXPERIMENT 23 (Page 1)

1. Acidic and Basic Solutions. (A)

Solution	Litmus	Phenolphthalein
1 M HCl, hydrochloric acid		
1 M H_2SO_4, sulfuric acid		
1 M NH_4^+, ammonium ion (from NH_4Cl)		
1 M H_3PO_4, phosphoric acid		
1 M $HC_2H_3O_2$, acetic acid		
1 M NaOH, sodium hydroxide		
1 M NH_3, ammonia		
1 M PO_4^{3-}, phosphate ion (from Na_3PO_4)		
1 M CO_3^{2-}, carbonate ion (from Na_2CO_3)		

2. Acid-Base Reactions

(B) (a) Temp. before mixing ________

Temp. after mixing ________

Color after mixing ________

(D) Write a balanced net-ionic equation for the acid-base reaction that took place.

(D) (b)

Litmus tests:

Balanced equation:

(E) (c)

Litmus tests:

Balanced equation:

(F) (d)

Litmus tests:

Balanced equation:

(G) (e)

Litmus tests:

Balanced equations:

(H) (f)

Solubility in water:

Litmus test:

Give a balanced equation to show what happened:

(I) (g)

Litmus test:

Result:

Give a balanced equation to show what happened:

REPORT SHEET

Name ________________

Lab Section ________________

Due Date ________________

EXPERIMENT 23 (Page 2)

(J) (h)

Litmus test:

Result:

Balanced equation:

(K) (i)

Description:

Give a balanced equation to show what happened:

(L) (j)

Solubility:

Description:

Give a balanced equation to show what happened:

23

QUESTIONS

Name ______________________________

Lab Section ______________________

Due Date __________________________

EXPERIMENT 23

1. Give definitions for:

(a) Acid

(b) Base

2. Give a balanced net-ionic equations for any acid-base reactions that occur upon mixing the following solutions. Refer to Table 23-2.

(a) A solution of potassium hydrogen sulfate, $KHSO_4$, is mixed with a solution of sodium hydroxide, NaOH.

(b) A solution of sodium hydrogen oxalate, $NaHC_2O_4$, is mixed with a solution of hydrochloric acid.

(c) A solution of sodium hydrogen carbonate, $NaHCO_3$, is mixed with a solution of sodium hydroxide, NaOH.

(d) A solution of potassium hydrogen carbonate, $KHCO_3$, is mixed with nitric acid.

24 Acid-Base Titration

Objective

To practice the titration method of analysis and to use the method to analyze a vinegar solution and a hydrochloric acid solution.

Discussion

A titration is the process of adding a measured volume of a solution of known concentration to a sample of another solution for purposes of determining the concentration of the second solution or the amount of some species in the solution. A species in the solution of known concentration reacts with another species in the unknown solution. The addition and measurement of the volume of the solution of known concentration is carried out by use of a buret. (See Fig. 24-1.) A titration is usually carried out by placing a measured sample of the unknown solution in a flask, filling the buret with the known solution (called the titrant) and then slowly delivering the titrant to the flask until the necessary amount has been added to the unknown solution.

The point at which the necessary amount has been added is called the end point of the titration. The end point is often detected by placing a small amount of a chemical called an indicator in the reaction flask. The indicator is chosen so that it will react with the titrant when the end point is reached. The reaction of the indicator produces a colored product; the appearance of the color signals the end point of the titration. Some indicators are colored to begin with and react at the end point to produce a different colored product, so the change in color indicates the end point. Other indicators change from a colorless form to a colored form at the end point.

Once the end point has been found, the volume of titrant used can be determined from the buret. Using this volume, the concentration of the titrant and the stoichiometric factor from the balanced equation, we can deduce the number of moles of species in the solution being titrated. If the molarity of the unknown solution is to be calculated, it is necessary to measure the volume of the original unknown solution before titrating. Then, the molarity can be found by dividing calculated number of moles by this volume. As an example, consider the following case. Determine the molarity of a hydrochloric acid solution if 30.21 mL of a 0.200 M sodium hydroxide solution is needed to titrate a 25.00 mL sample of the acid solution. First, the chemical reaction involved is

$$OH^-(aq) + H_3O^+(aq) \longrightarrow 2H_2O$$

Note that one mole of acid reacts with one mole of base. The number of moles of hydroxide ion needed to react can be found from the volume used and the molarity. The titration required 30.21 mL of 0.200 M NaOH. The number of moles of hydroxide ion is found by multiplying the volume of sodium hydroxide solution in liters by the molarity. The calculations are:

First the number of moles of OH^- used are found as the product of the volume and molarity (V_bM_b)

$$30.21\ \text{mL}\left(\frac{1\text{L}}{1000\ \text{mL}}\right)\left(\frac{0.200\ \text{moles OH}^-}{1\text{L}}\right) = 0.006042\ \text{moles OH}^-$$

Next the number of moles of hydronium ion is found by multiplying by the molar ratio obtained from the equation for the titration reaction.

$$0.006042\ \text{moles OH}^-\left(\frac{1\ \text{mole H}_3\text{O}^+}{1\ \text{mole OH}^-}\right) = 0.006042\ \text{moles H}_3\text{O}^+$$

Finally the molarity of the acid solution can be found by dividing the number of moles of hydronium ion by the original volume of the acid solution in liters.

$$\left(\frac{0.006042\ \text{moles H}_3\text{O}^+}{25.00\ \text{mL}}\right)\left(\frac{1000\ \text{mL}}{1\ \text{L}}\right) = 0.242\ \text{M H}_3\text{O}^+$$

If the acid and base in a titration react in a one-to-one molar ratio, as is the case in the above example, the calculations can be simplified by using the equation:

$$M_a = \frac{V_bM_b}{V_a}$$

where V_b is the volume of the base used in the titration, V_a is the volume of the original acid solution, M_b is the molarity of the base solution and M_a is the molarity of the acid solution. Using this simple equation for the example given above the molarity of the acid solution is:

$$M_a = \left(\frac{30.21\ \text{mL}}{25.00\ \text{mL}}\right) 0.200\ \text{M} = 0.242\ \text{M}$$

HOW TO USE THE BURET

A buret is a piece of glass tubing calibrated to deliver measured volumes of solution. A 50-mL buret is calibrated to deliver between 0 and 50 mL. Each etched line on the buret corresponds to a 0.1 mL increment of volume. By interpolation, the buret can be read to the nearest 0.01 mL. These instructions apply to a Mohr buret as pictured in Figure 24-1. If you use a buret with a glass or plastic stopcock, disregard the references to the rubber tubing and bead.

1. CLEANING

Place some water in the buret and allow it to run out through the tip by squeezing the rubber tube around the glass bead (or opening the stopcock). If you notice water drops adhering to the sides of the buret as it is drained, the buret needs cleaning. Clean the buret using tap water and a small amount of detergent. Use a buret brush to scrub the inside of the buret. Rinse the buret four or five times with tap water and allow some water to run out of the tip each time. Finally, rinse the buret three or four times with 10 mL portions of distilled water. By rotating the buret, allow the water to rinse the entire buret and be sure to rinse the tip by allowing some water to pass through.

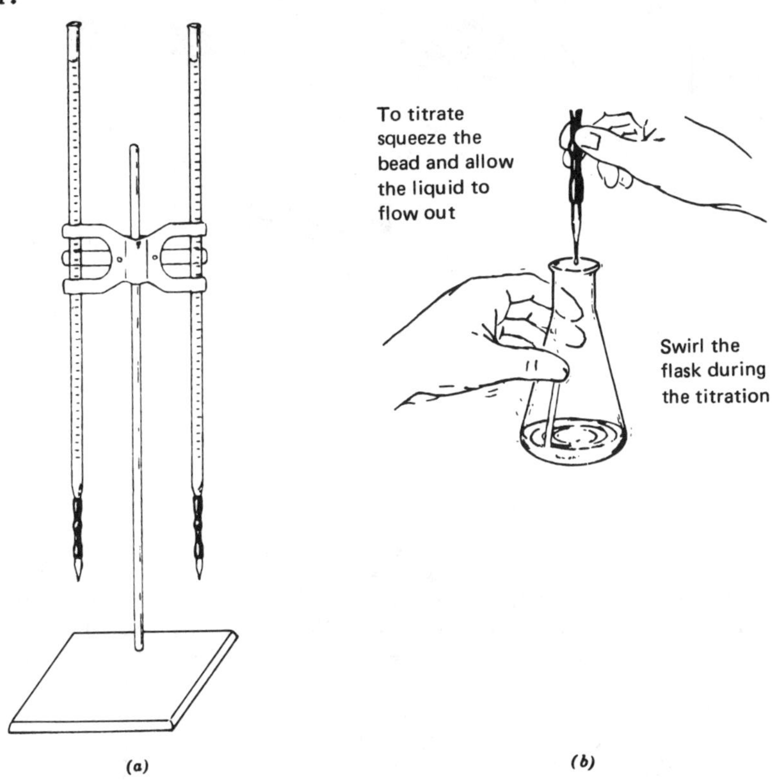

Fig. 24-1 Burets and the titration process.

2. FILLING

When filling a buret, try not to splash or spill any titrant. Clean up any spills. Use a paper towel to dry the outside of the buret. Rinse the clean buret two or three times with less than 5 mL portions of the solution with which it is to be filled. Be sure to rinse the tip each time. Fill the buret to a level just above the 0.00 mL mark. Fill the tip by bending the rubber to point the tip upward and then gently squeeze the tube to allow the liquid to displace the air (or open the stopcock and remove any air bubbles). Adjust the meniscus to a position somewhat below the 0.00 mL mark.

3. READING

The position of the bottom of the meniscus is read to the nearest 0.01 mL. The buret can be read directly to the nearest 0.1 mL, but you must interpolate to read to the nearest 0.01 mL. To interpolate, you imagine that the distance between lines is made up of 10 equally spaced parts and then locate in which of the parts the bottom of the meniscus is located. A buret reading card can be used to aid you in reading the position of the meniscus. (See Fig. 24-2.) When reading the buret, make sure your eye is level with the bottom of the meniscus.

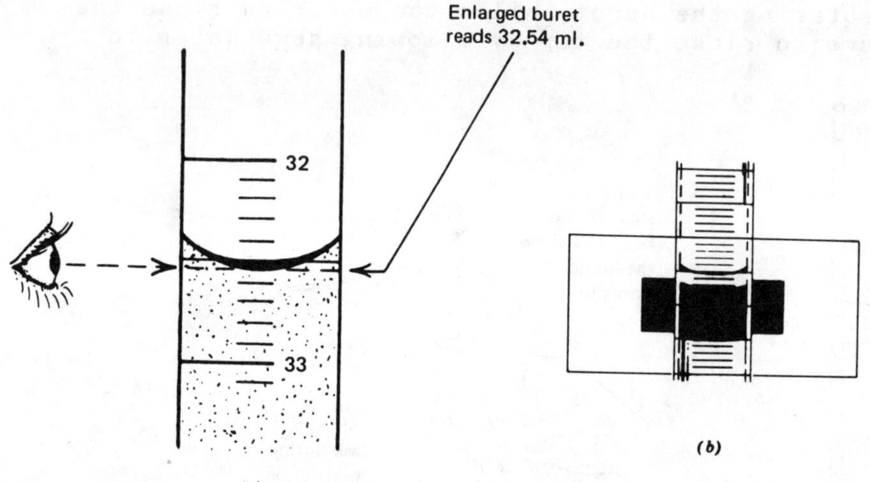

Fig. 24-2 Reading a buret. (a) Keep eye level with the bottom of the meniscus. (b) Use a buret reading card behind the buret to locate the bottom of the meniscus.

4. TITRATION METHOD

Be sure to record the initial and final buret readings when you do a titration. Place the sample of solution to be titrated in a flask. After recording the initial volume of the buret, allow the titrant to flow into the flask by pinching the rubber tube around the glass bead (or opening the stopcock). Let the titrant flow rapidly at first, and then add smaller and smaller volume increments as the end point is approached. When you are close to the end point, add one drop or less at a time. A fraction of a drop can be added by allowing a portion of a drop to form on the tip, touching the tip to the inside of the flask and then washing down the sides of the flask with a small amount of distilled water from a wash bottle. During the titration, mix the solutions in the flask by swirling but do not splash the solutions out of the flask. (See Fig. 24-1b.)

5. CLEANING

Drain the solution from the buret and rinse thoroughly with water. Remember to rinse the tip. Try not to spill any of the solutions on the desk or your clothing. If you do spill the solutions, clean them up. If any solutions spill on your clothing, tell your instructor. Burets can be stored by filling with distilled water and placing a cork or stopper in the top.

Laboratory Procedure

1. ANALYSIS OF ACETIC ACID IN VINEGAR

In this experiment you will titrate a vinegar solution using a standardized sodium hydroxide solution. You will need a clean, dry 100-mL beaker, a 125-mL flask, a 250-mL flask, a buret, a buret clamp and a ring stand.

Set up a buret on a ring stand as shown in Fig. 24-1 but use only one buret. Obtain about 100-mL of a standardized sodium hydroxide solution in a 100-mL beaker. Record the molarity of this solution. (A) Be very careful not to splash or spill this solution. Sodium hydroxide solutions are especially dangerous if splashed in your eyes. Always pour such solutions in a buret by removing the buret from the clamp and holding the buret in a piece of paper towel well below your eye level. Never raise the beaker of solution above your eye level. Rinse and fill your buret with the sodium hydroxide solution according to the instructions given in the discussion section.

You will find vinegar solution in a plastic bottle fitted with a 10.00 mL pipet. As demonstrated by your instructor, pipet a 10 mL sample of vinegar into a clean 250 mL flask. Add 2 or 3 drops of phenolphthalein indicator to the flask and dilute to about 50 mL volume using distilled water. Read the initial volume of the buret.(B) Titrate the acid solution sample with the sodium hydroxide solution until the end point is reached. As you add base, you will notice a slight pink color at the point at which the base solution enters the acid solution. As you approach the end point, you will see the entire solution flash pink. When this occurs, slow down the rate of addition of the base and carefully approach the end point. When the solution in the flask turns a very light pink and remains pink for one minute or longer you have reached the end point. The sides of the flask may be washed down with small amounts of distilled water at any time. Once the end point has been reached, read the final volume. (C)

Rinse the 250-mL flask with distilled water. Pipet another 10 mL of acid into the flask and titrate the acid with the base. You may have to refill the buret with some sodium hydroxide solution. Record the titration data in the table on the Report Sheet. Rinse the flask and carry out a third titration. When you are finished be sure to rinse out your buret with water and remember to rinse the tip of the buret so that it does not contain base solution.

Use the volumes of the acid and base solutions used from the data table for the calculations. If the acid-base reaction involved is

$$HC_2H_3O_2(aq) + OH^-(aq) \longrightarrow H_2O + C_2H_3O_2^-(aq)$$

calculate the molarity of acetic acid in vinegar for each titration. (D) If your results are within 10% of one another, calculate the average molarity. (E) If your results do not seem close enough, consult your instructor.

2. THE PERCENT BY MASS ACETIC ACID IN VINEGAR (OPTIONAL)

Assuming that the density of vinegar is 1.000 g/mL vinegar, it is possible to calculate the percent by mass acetic acid in vinegar. This is done by carrying out the following sequence of calculations. (AA represents acetic acid.)

$$\left(\frac{\text{moles AA}}{\text{1 L vinegar}}\right) \rightarrow \left(\frac{\text{moles AA}}{\text{1 mL vinegar}}\right) \rightarrow \left(\frac{\text{moles AA}}{\text{1 g vinegar}}\right) \rightarrow \left(\frac{\text{g AA}}{\text{g vinegar}}\right) 100 = \%\text{AA}$$

Using your experimental value for the molarity of acetic acid in vinegar, calculate the percent by mass acetic acid in vinegar. (F)

3. ANALYSIS OF AN UNKNOWN HYDROCHLORIC ACID SOLUTION (OPTIONAL)

In this experiment you will titrate an unknown acid solution using a standardized sodium hydroxide solution. You will need a clean, dry 100-mL beaker, a 125-mL flask, a 250-mL flask, a 20-mL pipet, a pipet bulb, a buret, a buret clamp and a ring stand.

Set up a buret on a ring stand as shown in Fig. 24-1 but use only one buret. Obtain about 100-mL of a standardized sodium hydroxide solution in a 100-mL beaker. Record the molarity of this solution. (G) Be very careful not to splash or spill this solution. Sodium hydroxide solutions are especially dangerous if splashed in your eyes. Always pour such solutions in a buret by removing the buret from the clamp and holding the buret in a piece of paper towel well below your eye level. Never raise the beaker of solution above your eye level. Rinse and fill your buret with the sodium hydroxide solution according to the instructions given in the discussion section.

Obtain a sample of unknown hydrochloric acid solution and record the number of the unknown. (H) Using the pipet and pipet bulb as demonstrated by your instructor, measure 20.00 mL of unknown solution and place it in a 250-mL flask. Add 2 or 3 drops of phenolphthalein indicator to the flask and dilute to about 50 mL volume using distilled water. Read the initial volume of the buret. (I) Titrate the acid solution sample with the sodium hydroxide solution until the end point is reached. As you add base, you will notice a slight pink color at the point at which the base solution enters the acid solution. As you approach the end point, you will see the entire solution flash pink. When this occurs, slow down the rate of addition of the base and carefully approach the end point. When the solution in the flask turns a very light pink and remains pink for one minute or longer you have reached the end point. The sides of the flask may be washed down with small amounts of distilled water at any time. Once the end point has been reached, read the final volume. (J)

Rinse the 250-mL flask with distilled water. Pipet another 20 mL of acid into the flask and titrate the acid with the base. You may have to refill the buret with some sodium hydroxide solution. Record the titration data in the table on the Report Sheet. Rinse the flask and carry out a third titration. When you are finished be sure to rinse out your buret with water and remember to rinse the tip of the buret so that it does not contain base solution.

Use your titration data to calculate the molarity of the unknown hydrochloric acid solution. (L) If your results are within about 10% of one another, calculate the average molarity. (M) If your results do not seem close enough, consult your instructor.

24

REPORT SHEET

Name _______________

Lab Section _______________

Due Date _______________

EXPERIMENT 24

1. Analysis of Acetic Acid in Vinegar

Data Table:

(A)	Molarity of NaOH solution	__________________________			
(C)	Final volume NaOH solution	________	________	________	________
(B)	Initial volume NaOH solution	________	________	________	________
	Volume NaOH solution used	________	________	________	________
	Volume vinegar	10.00 mL	10.00 mL	10.00 mL	10.00 mL
(D)	Calculated molarity of acetic acid in vinegar	________	________	________	________

(E) Average M acetic acid in vinegar

Give the setups for the molarity calculations below.

2. The Percent by Mass Acetic Acid in Vinegar (Optional)

(F) Give setup and calculated result.

3. Analysis of an Unknown Hydrochloric Acid Solution (Optional)

Data Table:

(G) Molarity of NaOH solution __________ (H) Unknown Number _________

(J)	Final volume NaOH solution	________	________	________	________
(I)	Initial volume NaOH solution	________	________	________	________
	Volume NaOH solution used	________	________	________	________
	Volume acid solution used	20.00 mL	20.00 mL	20.00 mL	20.00 mL
(L)	Calculated molarity acid solution	________	________	________	________

(M) Average M of acid solution

Give the setups for the molarity calculations below.

24

QUESTIONS

Name ______________________________

Lab Section ______________________

Due Date _________________________

EXPERIMENT 24

1. Explain the function and purpose of an indicator in a titration.

2. If you are titrating an acid with a base, tell how each of the following factors will affect the calculated molarity of the acid. That is, would the calculated molarity be greater than it should be, less than it should be, or not affected.

(a) The end point is exceeded by adding too much base.

(b) The volume of acid is measured incorrectly so that it is smaller than the value used in the calculations.

(c) Distilled water is used to wash down the sides of the flask during the titration.

3. A sample of vinegar is titrated with a sodium hydroxide solution to find the molarity of acetic acid. If 29.54 mL of a 0.435 M NaOH solution is required to titrate a 10.0 mL vinegar solution, what is the molarity of acetic acid in the vinegar?

24

QUESTIONS

Name [illegible]

Lab Section [illegible]

Due Date [illegible]

EXPERIMENT 24

1. [illegible]

2. If you [illegible] each of the following [illegible] would [illegible] the value [illegible]

(a) The end point is exceeded [illegible] each case.

(b) The volume [illegible] is measured [illegible] than the value used in the calculation.

(c) Distilled water is used [illegible] the titration.

3. [illegible]

25 Oxidation-Reduction Reactions

Objective

The purpose of this experiment is to observe some oxidation-reduction reactions occurring in aqueous solutions and to write equations for these reactions.

Discussion

Oxidation-reduction reactions are electron transfer reactions. Oxidation is the loss of electrons by some species and reduction is the gain of electrons by some species. In an oxidation-reduction or redox reaction, oxidation and reduction occur simultaneously. That is, in a redox reaction, one reactant is oxidized and another is reduced.

When a redox reaction occurs, the oxidation number of one element is increased and the oxidation number of another element is decreased. Oxidation is characterized by an increase in oxidation number and reduction is characterized by a decrease in oxidation number. When a reaction involves elements changing oxidation numbers, it is a redox reaction. An oxidizing agent is a species that can oxidize another species and a reducing agent is a species that can reduce another species. When a solution containing an oxidizing agent is added to a solution containing a reducing agent, a redox reaction may occur. However, not all oxidizing and reducing agents will react with one another. A redox reaction is represented by a net-ionic equation showing the oxidizing agent, the reducing agent and the products they form along with any other species needed to balance the equation.

Equations for redox reactions can be predicted using redox tables such as those given in Table 25-1 and 25-2. To predict a reaction we use the principle that an oxidizing agent will react with any reducing agent that is below it in the redox table. As an example, consider the reaction that occurs when a solution of potassium permanganate (K^+ + MnO_4^-) is mixed with a solution of sodium bromide (Na^+ + Br^-) in dilute sulfuric acid. The acid serves as a source of H_3O^+. The products of the reaction are manganese(II) ion, Mn^{2+}, and Br_2. From Table 25-1 we can see that it is expected that an acidified solution of MnO_4^- will react as an oxidizing agent and oxidize the reducing agent, Br^-. To write the equation for the reaction, first write the MnO_4^- half reaction as it is found in the table and, below this, write the Br^- reaction in the reverse of that which is in the table.

$$MnO_4^- + 8H_3O_+ + 5e^- \longrightarrow 12H_2O + Mn^{2+}$$

$$2Br^- \longrightarrow Br_2 + 2e^-$$

To balance the electrons lost with the electrons gained, multiply the MnO_4^- half reaction by 2 and the Br^- half reaction by 5. Then add the two half reactions for the complete equation as shown on the next page.

$2MnO_4^- + 16H_3O^+ + 10e^-$		$24H_2O + 2Mn^{2+}$
$10Br^-$		$5Br_2 + 10e^-$
$2MnO_4^- + 16H_3O^+ + 10Br^-$		$5Br_2 + 24H_2O + 2Mn^{2+}$

Table 25-1 A Redox Table Including Some Common Oxidizing and Reducing Agents

Oxidizing Agents		Reducing Agents
$H_2O_2 + 2H_3O^+ + 2e^-$	$\rightleftarrows$	$4H_2O$
$MnO_4^- + 8H_3O^+ + 5e^-$	$\rightleftarrows$	$Mn^{2+} + 12H_2O$
$Cl_2(g) + 2e^-$	$\rightleftarrows$	$2Cl^-$
$Cr_2O_7^{2-} + 14H_3O^+ + 6e^-$	$\rightleftarrows$	$2Cr^{3+} + 21H_2O$
$Br_2(\ell) + 2e^-$	$\rightleftarrows$	$2Br^-$
$NO_3^- + 4H_3O^+ + 3e^-$	$\rightleftarrows$	$NO(g) + 6H_2O$
$Fe^{3+} + 1e^-$	$\rightleftarrows$	Fe^{2+}
$I_2(s) + 2e^-$	$\rightleftarrows$	$2I^-$
$2CO_2(g) + 2H_3O^+ + 2e^-$	$\rightleftarrows$	$H_2C_2O_4 + 2H_2O$

Table 25-2 A Redox Table Including Common Metals

Oxidizing Agents		Reducing Agents	
$Ag^+ + 2e^-$	$\rightleftarrows$	$Ag(s)$	
$Hg^{2+} + 1e^-$	$\rightleftarrows$	$Hg(\ell)$	
$Fe^{3+} + 1e^-$	$\rightleftarrows$	Fe^{2+}	
$Cu^{2+} + 2e^-$	$\rightleftarrows$	$Cu(s)$	
$2H_3O^+ + 2e^-$	$\rightleftarrows$	$H_2(g) + 2H_2O$	
$Pb^{2+} + 2e^-$	$\rightleftarrows$	$Pb(s)$	- - - - - - - - - - - - - - - -
$Ni^{2+} + 2e^-$	$\rightleftarrows$	$Ni(s)$	These metals dissolve in acid solutions.
$Co^{2+} + 2e^-$	$\rightleftarrows$	$Co(s)$	
$Cd^{2+} + 2e^-$	$\rightleftarrows$	$Cd(s)$	
$Fe^{2+} + 2e^-$	$\rightleftarrows$	$Fe(s)$	
$Zn^{2+} + 2e^-$	$\rightleftarrows$	$Zn(s)$	
$2H_2O + 2e^-$	$\rightleftarrows$	$H_2(g) + 2OH^-$	
$Al^{3+} + 3e^-$	$\rightleftarrows$	$Al(s)$	
$Mg^{2+} + 2e^-$	$\rightleftarrows$	$Mg(s)$	- - - - - - - - - - - - - - - -
$Na^+ + 1e^-$	$\rightleftarrows$	$Na(s)$	These metals react spontaneously in water.
$Ca^{2+} + 2e^-$	$\rightleftarrows$	$Ca(s)$	
$K^+ + 1e^-$	$\rightleftarrows$	$K(s)$	

How do we know when a reaction has occurred in a solution? To "see" a reaction, some observable and noticeable change must accompany the reaction. Consider the following examples that illustrate typical changes that accompany redox reactions.

1. The consumption or dissolving of a solid:

$$Zn(s) + 2H_3O^+ \longrightarrow Zn^{2+} + 2H_2O$$

Solid zinc metal dissolves in acid.

2. The formation of a gas:

$$H_2O_2 + 2H_3O^+ + 2Cl^- \longrightarrow Cl_2(g) + 4H_2O$$

Chlorine gas is produced. (DANGER: Do not try this reaction.)

3. The formation of a solid:

$$2Al + 3Cu^{2+} \longrightarrow 3Cu(s) + 2Al^{3+}$$

Solid copper is formed as solid aluminum dissolves.

4. A color change (a colored reactant forms a colored product, a colored reactant is consumed or a colored product is formed):

$$8H_3O^+ + Cr_2O_7^{2-} + 3H_2C_2O_4 \longrightarrow 2Cr^{3+} + 6CO_2 + 15H_2O$$

Yellow-orange $Cr_2O_7^{2-}$ changes to green Cr^{3+}.

Of course reactions other than redox can involve changes. Such changes are indications of chemical reactions. However, many redox reactions will display one or more of the changes describe here and we can use the changes as an indication of a reaction.

Laboratory Procedure

In the following exercises space is provided for your observations and equations. Summarize your observations and give balanced equations on the Report Sheet.

1. OBSERVING REDOX REACTIONS

In this section, a redox reaction will be observed. Describe any evidence of a reaction and write an equation for the reaction.

Place a small piece of aluminum foil in a small test tube. Add about 2 mL of a 6 M hydrochloric acid solution to the tube. Allow the reaction to occur and observe any changes that you see taking place. Record these observations. (A)

As observed, aluminum dissolves, a gas is produced, and the solution turns cloudy and then clears. Since we started with solid aluminum and added hydrochloric acid, the formulas of the species that were mixed are Al and H_3O^+ + Cl^-.

To explain the reaction that occurred, refer to Table 25-2. Note that aluminum metal is below hydronium ion (H_3O^+) in the table. Aluminum will react with hydronium ion. To write the equation, use the half reaction of H_3O^+ from the table and the reverse of the aluminum half reaction from the table. (B)

The predicted reaction is the sum of these two half reactions. (If necessary, multiply each by the proper number to make the number of electrons lost equal the number gained.) Write the balanced equation for the reaction. (C)

2. SOME TYPICAL OXIDATION-REDUCTION REACTIONS

For each of the following, mix the solutions in a small test tube and stir with a clean glass rod. The volumes used need only be approximate. Record any changes and evidence of any reaction. Do not hurry, but rather allow time for any slower reactions to take place. Using the formulas of the species that are mixed the half reactions can be deduced from a redox table and a balanced net-ionic equation can be written for any reaction that occurred. Refer to any redox tables for half reactions.

(a) Place 1 mL of a 0.1 M sodium dichromate solution in a test tube and add 1 mL of 6M hydrochloric acid. Now add some solid iron(II) ammonium sulfate (a sample about the size of a pencil eraser). Stir and observe any changes. (D) The species mixed are Na^+ + $Cr_2O_7^{2-}$ + H_3O^+ + Cl^- and Fe^{2+} + NH_4^+ + SO_4^{2-}. (Hint: The iron(II) ion is involved in the reaction but the ammonium and sulfate ions are not involved.)

(b) Place 1 mL of a 0.1 M potassium permanganate solution in a test tube and add 1 mL of 1 M sulfuric acid. Now add about 1 mL of 1 M oxalic acid and look for evidence of a gas being formed. (E) The species mixed are K^+ + MnO_4^- + H_3O^+ + HSO_4^- and $H_2C_2O_4$.

(c) Place 1 mL of 3% hydrogen peroxide solution in a test tube and add 1 mL of 1 M sulfuric acid. Work in the fume hood. Now add 1 mL of 0.1 M potassium iodide solution. Stir and observe any evidence of a reaction. (F) The species mixed are H_2O_2 + H_3O^+ + HSO_4^- and K^+ + I^-.

(d) Pour about 0.5 mL of a 0.1 M silver nitrate solution in a test tube. Obtain a small length of copper wire and place it in the test tube so that the end dips into the solution. Allow the wire to remain in the solution for a few minutes. Remove the wire and inspect the end of the wire. Return the length of wire to the stockroom or your instructor. (G) The species mixed are Ag^{+} + NO_3^{-} and Cu.

(e) Pour about 1 mL of a 0.1 M copper(II) nitrate solution into a test tube. Now drop in a small piece of solid zinc metal. Describe any changes. (H) The species mixed are Cu^{2+} + NO_3^{-} and Zn.

(f) Pour 1 mL of a 1 M hydrochloric acid solution into a test tube. Drop in a small piece of magnesium metal. Be careful of any spattering. (I) The species mixed are H_3O^{+} + Cl^{-} and Mg.

(g) Obtain a piece of calcium metal and, if necessary, rub it on a piece of sandpaper to remove some oxide coating and expose the metal. Add the metal to 2 mL of distilled water in a test tube. Be careful that it does not bubble over. The species mixed are Ca and H_2O. (Hint: One of the products of the reaction is solid, white calcium hydroxide, $Ca(OH)_2$. (J)

3. THE HALOGENS AND HALIDE IONS

The halogens include fluorine, chlorine, bromine, and iodine. The halide ions are fluoride ion, chloride ion, bromide ion, and iodide ion. Elemental halogens are different from the halide ions. The halogens in the elemental form occur in diatomic molecular form and the halide ions occur as components of ionic compounds. Give the formulas of the diatomic molecules and ions of the various halogens. (K)

Element	Symbol	Molecular Formula	Halide Ion Formula
Fluorine	F		
Chlorine	Cl		
Bromine	Br		
Iodine	I		

The elemental halogens occur in various colors. Fluorine is a pale-yellow gas but it is so chemically reactive that fluorine samples are seldom used in the laboratory. Samples of the other halogens are on display in the laboratory along with samples of solutions of the halogens in hexane. Observe the samples and record the colors on the next page. (L)

Element	Color of Element	Color of Hexane Solution
Chlorine		
Bromine		
Iodine		

Some typical ionic compounds containing the halide ions and water solutions of these compounds are on display in the laboratory. Observe the display and fill in the following table: (M)

Formula of Compound	Color and Form of Compound	Color of Water Solution
____________		
____________		
____________		

What do you conclude about the color of the halide ions? (N)

4. PREPARING HALOGENS

In this exercise, samples of the halogens chlorine, bromine, and iodine are to be prepared. The reactions used to prepare the halogens will be carried out in water. Some hexane, which is not soluble in water, will be used to extract the halogens as they are formed. That is, the halogens are soluble in hexane, so if we have hexane in contact with the reaction mixture, the halogen formed will be captured in the hexane layer. Hexane is not part of any chemical reaction but just a convenient solvent for the halogens. Halogens are toxic and hexane is flammable, so we will work in the fume hood for all parts of this exercise. You may use approximate volumes of solutions.

(a) Preparing Chlorine - Place 1 mL of a 0.1 M $KMnO_4$ solution in a test tube. Add about 2 mL of hexane. The hexane will float on top of the water. Add 1 mL of 6 M hydrochloric acid. Cork the tube and mix by repeatedly inverting the tube for a few minutes. Describe any changes and give a balanced net-ionic equation for the formation of chlorine. (See Table 25-1.) (O) Save the contents of the tube for part b.

(b) Preparing Bromine - The chlorine formed in part a will be used to form bromine from bromide ion. You will need a clean test tube and an eye dropper. Remove the cork from the tube of part a. Use your wash bottle to carefully add water to the tube to raise the level of the hexane so that it can be removed. Use an eye dropper to transfer the hexane layer to a clean test tube. Add 1 mL of 0.1 M KBr to the test tube. Cork the tube and mix by repeatedly inverting the tube for a few minutes. Describe any changes and give a balanced net-ionic equation for the formation of bromine. (See Table 25-1.) (P) Save the contents of the tube for part c.

(c) Preparing Iodine - The bromine formed in part b will be used to form iodine from iodide ion. You will need a clean test tube and an eye dropper. Remove the cork from the tube of part b. Use your wash bottle to carefully add water to the tube to raise the level of the hexane so that it can be removed. Use an eye dropper to transfer the hexane layer to a clean test tube. Add 1 mL of 0.1 M KI to the test tube. Cork the tube and mix by repeatedly inverting the tube for a few minutes. Describe any changes and give a balanced net-ionic equation for the formation of iodine. (See Table 25-1.) (Q) chlorine. (See Table 25-1, c.) (e) Save the contents of the tube for part b.

(b) Preparing Bromine — The chlorine formed in part a will be used to form bromine from bromide ion. You will need a clean test tube and an eye dropper. Remove the cork from the tube of part a. Use your wash bottle to carefully add water to the tube to raise the level of the hexane so that it can be removed. Use an eye dropper to transfer the hexane layer to a clean test tube. Add 1 mL of 0.1 M KBr to the test tube. Cork the tube and mix by repeatedly inverting the tube for a few minutes. Describe any changes and give a balanced net-ionic equation for the formation of bromine. (See Table 25-1, d.) (f) Save the contents of the tube for part c.

(c) Preparing Iodine — The bromine formed in part b will be used to form iodine from iodide ion. You will need a clean test tube and an eye dropper. Remove the cork from the tube of part b. Use your wash bottle to carefully add water to the tube to raise the level of the hexane so that it can be removed. Use an eye dropper to transfer the hexane layer to a clean test tube. Add 1 mL of 0.1 M KI to the test tube. Cork the tube and mix by repeatedly inverting the tube for a few minutes. Describe any changes and give a balanced net-ionic equation for the formation of iodine. (See Table 25-1, e.) (g)

25

REPORT SHEET

Name _______________

Lab Section _______________

Due Date _______________

EXPERIMENT 25 (Page 1)

1. Observing Redox Reactions

(A)

(B)

(C)

2. Some Typical Oxidation-Reduction Reactions

(a) (D)

(b) (E)

(c) (F)

(d) (G)

(e) (H)

(f) (I)

(g) (J)

3. The Halogens and Halide Ions
(K)

Element	Symbol	Molecular Formula	Halide Ion Formula
Fluorine	F		
Chlorine	Cl		
Bromine	Br		
Iodine	I		

(L)

Element	Color of Element	Color of Hexane Solution
Chlorine		
Bromine		
Iodine		

(M)

Formula of Compound	Color and Form of Compound	Color of Water Solution
----------		

What do you conclude about the color of the halide ions? (N)

EXPERIMENT 25

Name _______________

Lab Section _______________

Due Date _______________

EXPERIMENT 25 (Page 2)

4. Preparing Halogens

(a) Preparing Chlorine (O)

(b) Preparing Bromine (P)

(c) Preparing Iodine (Q)

25

QUESTIONS

Name________________________________

Lab Section ______________________

Due Date __________________________

EXPERIMENT 25

1. Give definitions for the following terms:

(a) oxidation

(b) reduction

2. Using Tables 25-1 and 25-2, give the balanced net-ionic equations for any reactions that may occur when the following solutions are mixed.

(a) A $Na_2Cr_2O_7$ solution made acidic with sulfuric acid is added to a solution of oxalic acid, $H_2C_2O_4$.

(b) A piece of cobalt metal is added to hydrochloric acid.

(c) A piece of nickel metal is added to a $Hg(NO_3)_2$ solution.

3. Using your observations and reactions in part 4 of experiment 25, explain the following statement. Any halogen will displace a halide ion which is below it in the periodic table.

QUESTIONS

Name ____________________

Lab Section ____________________

Due Date ____________________

EXPERIMENT 25

1. Give definitions for the following terms:

(a) oxidation

(b) reduction

2. Using Tables 25-1 and 25-2, give the balanced net ionic equations for any reactions that may occur when the following solutions are mixed.

(a) A $Na_2Cr_2O_7$ solution made acidic with sulfuric acid is added to a solution of oxalic acid, $H_2C_2O_4$.

(b) A piece of cobalt metal is added to hydrochloric acid.

(c) A piece of nickel metal is added to a $Hg(NO_3)_2$ solution.

3. Using your observations and conclusions in part 4 of experiment 25, explain the following statement: Any halogen will displace a halide ion which is below it in the periodic table.

26 The Preparation of Aspirin

Objective

The purpose of this experiment is to prepare some aspirin and determine the percent yield of the reaction.

Discussion

Aspirin or acetylsalicylic acid is the most common medicinal drug in use today. Aspirin is an analgesic or pain reliever and an antipyretic or fever reducer. Aspirin is often recommended for minor pain but it does have side effects when used excessively or by persons who are susceptible. Aspirin is known to cause dizziness, nausea, upset stomach, and bleeding of the stomach. Before using aspirin to treat children suffering from flu, a doctor or pharmacist should be consulted.

Aspirin can be made by reacting salicylic acid and acetic anhydride as shown by the following equation:

$C_7H_6O_3$	+	$C_4H_6O_3$	$\longrightarrow$	$C_9H_8O_4$	+	$HC_2H_3O_2$ 2
salicylic acid		acetic anhydride		aspirin		acetic acid

In this experiment, aspirin is to be prepared by reacting salicylic acid and acetic anhydride. Aspirin is quite insoluble in water so it can be crystallized in water and separated by filtration. Specific amounts of the reactants will be mixed to produce aspirin. The aspirin produced will be isolated and purified. However, it will not be pure enough for you to use.

Laboratory Procedure

CAUTION: Some of the chemicals used in this experiment are dangerous or very irritating to eyes and skin. Use them with care as instructed. Wear safety glasses and avoid breathing any fumes by keeping the flask away from your nose.

For the experiment you will need a 125-mL Erlenmeyer flask, a 400-mL beaker, two 100-mL beakers, a Büchner funnel (See Fig. 26-1), a filter flask and a one-holed rubber stopper to fit the 125-mL flask.

Obtain about 10.3 g of solid salicylic acid. Weigh a clean, dry 125-mL flask to 0.01 g. (B) Transfer the sample of salicylic acid to the flask using a piece of creased paper. Weigh the flask plus the salicylic acid to 0.01 g. (A) Subtract the masses to find the mass of salicylic acid used. (C)

Fill a 400-mL beaker about two-thirds full of water and bring it to a boil by placing it on a wire gauze on a ring over a Bunsen burner flame. Take the 125-mL flask to the fume hood. Measure out about 13 mL of acetic anhydride in a graduated cylinder to the nearest 0.1 mL. (D) (CAUTION: Acetic anhydride is very irritating. Do not breath the vapors. If you spill any on your skin, wash it off with large amounts of water.) Add the acetic anhydride to the flask. Keep the flask in the hood and add concentrated sulfuric acid from the dropper bottle provided. (CAUTION: Sulfuric acid is dangerous to the skin and eyes. If you spill or splash any on you, wash it off immediately using large amounts of water.) Carefully add 15 drops of concentrated sulfuric acid to the flask, swirling the flask to mix after the addition of each 2 to 3 drops of acid.

Place the one-holed stopper in the flask. Swirl to mix all of the chemicals. Place the flask in the beaker of gently boiling water. Control the boiling of the water by adjusting the burner flame. Heat the contents of the flask for 15 minutes. If any solid remains at this time, swirl the flask and continue heating for 10 minutes.

Remove the flask and cool by placing it in running tap water. Cool further by removing the stopper and adding about 30 mL of ice water and placing the flask in a beaker of crushed ice. Aspirin crystals should form at this time. If crystals are slow to appear, it may help to scratch the inside of the flask with a stirring rod. Leave the flask in the beaker of ice until it appears that no more crystals are forming.

Set up a Büchner funnel and filter flask as shown in Figure 26-1 and connect the flask to an aspirator.

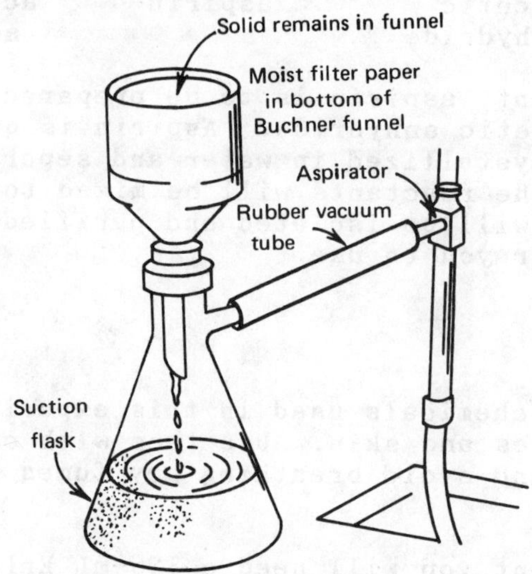

Fig. 26-1 A Büchner funnel and filter flask.

This apparatus will be used to filter the aspirin by suction filtration. Place a piece of filter paper in the funnel and wet the paper. Turn on the suction and filter the aspirin by transferring the contents of the flask to the filter. Use a spatula and a stream of water from your wash bottle to transfer the solid aspirin to the funnel. When the filtration is complete, turn off the suction and lift the filter paper with the aspirin out of the funnel. Transfer all of the aspirin to a 100-mL beaker and scrape any aspirin off of the filter paper with a spatula.

The aspirin is not pure since it will contain some of the reactants. It can be purified by recrystallizing it from alcohol. (CAUTION: Even after purification, the aspirin will not be pure enough to use so do not attempt to use it.)

Take the beaker of impure aspirin to the hood. Add about 25 mL of pure ethyl alcohol. (CAUTION: Ethyl alcohol is very flammable so do not work near a Bunsen flame. Make sure no burners are in the fume hood.) Stir the alcohol and aspirin mixture. If the crystals do not all dissolve, bring a beaker of hot water to the hood, hold the beaker in the water and stir the contents to dissolve the aspirin.

After all of the solid has dissolved, add about 70 mL of distilled water and place the beaker in a slightly larger beaker of ice. When crystallization again appears to be complete filter the aspirin by suction filtration. Suck as much liquid as possible from the aspirin before turning off the suction.

Remove the filter paper with the aspirin and spread the product on a piece of paper towel. Dry the aspirin by placing the paper towel under a heat lamp or by letting it sit in your drawer for a few days.

To find the mass of the aspirin produced, weigh a small beaker to 0.01 g. (F) Place the dry aspirin in the beaker and weigh to 0.01 g. (E) Subtract for the mass of aspirin. (G)

From the equation for the reaction between salicylic acid and acetic anhydride and the masses of the reactants, calculate the expected amount of aspirin. To find the mass of acetic anhydride from the volume used you will need the density of acetic anhydride. Look up the density in a reference book or consult your instructor. This is a limiting reactant question so determine which of the reactants will form the lesser amount of aspirin. This will be the theoretical yield. (H) The actual yield of aspirin will be less than the theoretical yield because of incomplete reactions and loss of product during filtration. The actual yield is the mass of aspirin actually formed in the experiment. The percent yield is the ratio of the actual yield to the theoretical yield expressed as a percent. (I)

$$\%\ \text{yield} = \left(\frac{\text{actual yield}}{\text{theoretical yield}}\right) 100$$

26

REPORT SHEET

Name ______________

Lab Section ______________

Due Date ______________

EXPERIMENT 26

Preparation of Aspirin

(A) Mass of flask and salicylic acid ________________

(B) Mass of flask ________________

(C) Mass of salicylic acid ________________

(D) Volume of acetic anhydride ________________

Density of acetic anhydride ________________

Mass of acetic anhydride ________________

(E) Mass beaker and aspirin ________________

(F) Mass of beaker ________________

(G) Mass of aspirin (actual yield) ________________

(H) Theoretical yield (Show setup and results of calculations.)

(I) Percent yield

26

QUESTIONS

Name ________________________

Lab Section ________________

Due Date ___________________

EXPERIMENT 26

1. Use the reaction of salicylic acid and acetic anhydride to determine the number of grams of aspirin which could be formed from 525 g of salicylic acid.

2. Use the equation for the formation of aspirin from salicylic acid to answer the following question. A typical aspirin tablet contains 5 grains of aspirin. There are 15.4 grains per gram. How many grams of salicylic acid are needed to form 5 grains of aspirin?

3. Over 30 million pounds of aspirin are manufactured in the United States each year. If a typical aspirin tablet contains 5 grains of aspirin and there are 15.4 grains per gram, how many aspirin tablets could be made from 30 billion pounds of aspirin?

Appendix 1 The Vapor Pressures of Water

THE VAPOR PRESSURES OF WATER

Temperature °C	Pressure Torr	Temperature °C	Pressure Torr
0	4.58	35	42.2
5	6.54	40	55.3
10	9.21	45	71.9
15	12.79	50	92.5
20	17.54	55	118.0
21	18.65	60	149.4
22	19.83	65	187.5
23	21.1	70	234.
24	22.4	75	289.
25	23.8	80	355.
26	25.2	85	434.
27	26.7	90	526.
28	28.3	95	634.
29	30.0	100	760.
30	31.8		

Appendix 2 Laboratory Acids and Bases

COMMON LABORATORY ACIDS AND BASES

Solution		Molarity
Concentrated sulfuric acid	conc. H_2SO_4	18 M H_2SO_4
Concentrated hydrochloric acid	conc. HCl	12 M HCl
Concentrated nitric acid	conc. HNO_3	16 M HNO_3
Concentrated acetic acid	conc. $HC_2H_3O_2$	17 M $HC_2H_3O_2$
Concentrated phosphoric acid	conc. H_3PO_4	15 M H_3PO_4
Concentrated ammonia (sometimes labeled as aqua ammonia or ammonium hydroxide)	conc. NH_3	15 M NH_3
Dilute sulfuric acid	dil. H_2SO_4	3 M H_2SO_4
Dilute hydrochloric acid	dil. HCl	6 M HCl
Dilute nitric acid	dil. HNO_3	6 M HNO_3
Dilute acetic acid	dil. $HC_2H_3O_2$	6 M $HC_2H_3O_2$
Dilute ammonia (sometimes labeled as ammonium hydroxide)	dil. NH_3	6 M NH_3
Dilute sodium hydroxide	dil. NaOH	6 M NaOH

Appendix 2 Laboratory Acids and Bases

COMMON LABORATORY ACIDS AND BASES

Solution		Molarity
Concentrated sulfuric acid	conc. H_2SO_4	18 M H_2SO_4
Concentrated hydrochloric acid	conc. HCl	12 M HCl
Concentrated nitric acid	conc. HNO_3	16 M HNO_3
Concentrated acetic acid	conc. $HC_2H_3O_2$	17 M $HC_2H_3O_2$
Concentrated phosphoric acid	conc. H_3PO_4	15 M H_3PO_4
Concentrated ammonia (sometimes labeled as aqua ammonia or ammonium hydroxide)	conc. NH_3	15 M NH_3
Dilute sulfuric acid	dil. H_2SO_4	3 M H_2SO_4
Dilute hydrochloric acid	dil. HCl	6 M HCl
Dilute nitric acid	dil. HNO_3	6 M HNO_3
Dilute acetic acid	dil. $HC_2H_3O_2$	6 M $HC_2H_3O_2$
Dilute ammonia (sometimes labeled as ammonium hydroxide)	dil. NH_3	6 M NH_3
Dilute sodium hydroxide	dil. NaOH	1.0 M NaOH

Appendix 3 Graphing

Sometimes we collect data in the laboratory that represents how one property depends upon another. For example, we may observe how the volume of a gas sample changes with the temperature. The observed properties are called variables and we measure various values of these variables by experiment. Normally in an experiment one variable is changed in a controlled fashion and corresponding values of the other variable are measured. For instance, we may change the temperature of a gas sample and measure the corresponding volumes. The variable for which the changes are controlled is called the independent variable. The other variable, that changes with the independent variable, is called the dependent variable. The temperature of the gas sample is the independent variable and the volume is the dependent variable.

To show how the variables change with respect to one another it is possible to graph or plot the data on graph paper. The experimental data consists of a set of independent variable values and the corresponding dependent variable values. A data point is any two related variable values. Thus, the data consists of a set of data points. The plotting of data on graph paper involves labeling the graph, locating each point of the graph and connecting the points with a smooth curve. A straight line is drawn through the points if the data appears to fit a straight line or we know that the data should be represented by a straight line.

A piece of graph paper contains a square grid with a horizontal axis called the x-axis and a vertical axis called the y-axis. Normally, the independent variable is plotted on the x-axis and the dependent variable on the y-axis. When data is graphed in this way it is said that the dependent variable is plotted versus the independent variable. On a piece of graph paper the long or short side of the grid can be used for either axes as needed. That is, the long side can be used for the x-axis or the short side can be used to best fit the data. If it is not obvious which variable is the independent variable any side can be used as desired.

The general steps used in graphing data are:

1. Note the range of the variables. The range of each variable extends from the lowest value to the highest value.

2. Note the number of divisions on the graph paper. Some paper has a set of large divisions which are divided into smaller divisions.

3. Depending on the ranges of the variables decide which axis on the paper is to be the x-axis and which is to be the y-axis.

4. Using the ranges of the variables decide how the lines on the axes are to be numbered. Mark each axis with an appropriate sequence of numbers. Not all division need be numbered.

5. Label the axes of the graph with appropriate titles and show the units of the variables being plotted.

6. Locate each value of the independent variable on the x-axis and, then, move along the y-axis to find the location of the the corresponding value of the dependent variable. You need to pay close attention to the value of each division on the graph so that the values are located correctly. Mark a fine pencil or pen dot at the location of each point on the graph paper. Draw a small circle around each point on the graph to emphasize its location.

7. Once all of the points have been located connect the points with a smooth curve that best fits the points. If the variables seem to fall along a straight line or if you know that the variables should correspond to a straight line, draw the best straight line through the points. By best straight line is meant a line that lies as close as possible to all of the points. Use a ruler to draw a straight line. Some of the points may be located above or below the line but try to make the distances between the points and the line to be as small as possible.

To illustrate graphing, consider plotting some experimental data. Observation of the number of cricket chirps per minute at various temperatures produced the following results (See Croxton and Cowden, Applied General Statistics, Prentice-Hall, Englewood, N.J., 1940).

Temperature	45 °F	48 °F	50 °F	53 °F	56 °F	62 °F	64 °F	66 °F
Chirps/min	32	42	50	61	72	94	101	109

A plot of this data will show the relationship between chirps/min and temperature.

The first step in constructing a graph is to obtain a piece of graph paper. Graph paper comes in a variety of forms. This discussion will be based upon a piece of graph paper like that shown in Fig. A-1. The first step is to look at the range of the data and the divisions on the graph paper to decide how to number the divisions to fit the data. The temperature data ranges from 45 to 66 °F to give a range of about 30. The chirps per minute data ranges from 32 to 109 to give a range of about 80.

The long side of the graph paper is divided into nine large division each of which is divided into ten parts for a total of 90 divisions. If we divide the chirps/min range by this number of divisons we get 80/90 = 0.89 chirps/min per division. This is close to 1 chirp/min per division so it is possible to use the small divisions to each represent 1 chirps/min and each large division will then represent 10 chirps/min. Starting with 20 chirps/min and numbering to 110 chirps/min the y-axis can be numbered as as shown in Fig. A-1.

In general, to decide on the numbering of an axis the approximate range of the data can be divided by the number of divisions on the axis. Then the number of data units per division to be used is determined by rounding off this ratio to a convenient value. Typically the rounded value should represent some whole number multiple or fraction of data units per division. In our example, it worked out so that each division on the y-axis can represent one data unit or 1 chirp/min.

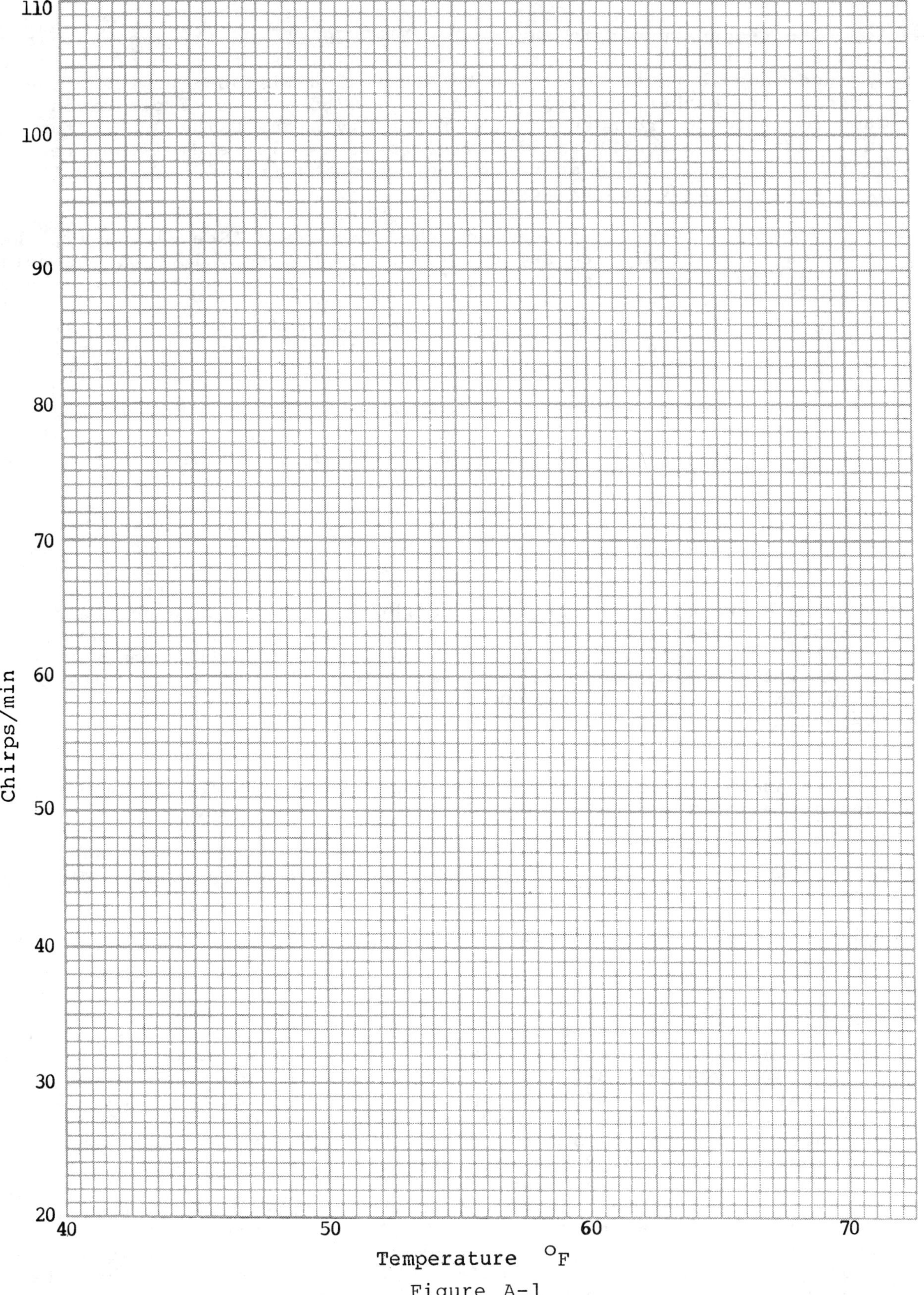
110
100
90
80
70
60
50
40
30
20
Chirps/min
40
50
60
70
Temperature °F

Figure A-1

The temperature data ranges roughly 30 °F. On the graph paper the x-axis is divided into 6 and 1/2 large divisions and each of these is divided into 10 parts to give a total of 65 divisions. Dividing the range by the number of divisions gives 30/65 = 0.46 °F per division. This ratio can be rounded to 0.5 °F per division. Using 0.5 °F for each small division makes 20 small divisions or two large divisions represent 10 °F. Using this approach and numbering from 40 °F to 70 °F the x-axis can be numbered as shown in Fig. A-1.

Once we number the lines and label the axes (See Fig. A-1), the data can be plotted. Each set of two corresponding values of the data will be a point on the graph. The first point is 32 chirps per minute at 45 °F. To locate this point move along the x-axis to 45, then move straight up along the y-axis to 32. Keep in mind that each division on the x-axis represents 0.5 °F. Make a dot at this point as shown in Fig. A-2. Put a small circle around the point to emphasize its location. The next point is 42 chirps/min at 48 °F. Move along the x-axis to 48 and, then, move straight up along the y-axis to 42. Make a dot at this point and circle it. The process is repeated for each point. Fig. A-2 shows all of the data points plotted except for 101 chirps/min at 64 °F. For practice, plot this point on the graph.

After all of the points have been plotted, we normally try to connect the points with as smooth a curve as possible. Such a curve may be a straight line or a curved line which ever seems to fit the data. The graph of the data shown in Fig. A-2 seems to fit a straight line. Fig. A-3 shows the graph with a straight line drawn through the data points. It appears, over the temperature range of the data, that a linear or straight line relationship exists between the temperature and the chirps per minute. It can be stated that the number of cricket chirps per minute is directly proportional to the Fahrenheit temperature.

As another example of graphing consider the following data which represents the way in which the density of a gas changes with the temperature.

Temperature (°C)	0	51	102	148	204	250	301	349	399	453	504
Density (g/L)	1.29	1.08	0.94	0.83	0.74	0.67	0.61	0.57	0.53	0.49	0.45

Considering the range of the data (about 0 to 600 and about 0 to 1.3), the divisions on the graph paper can be numbered as shown in Fig. A-4. Once the axes are numbered and labeled, the points can be plotted and a smooth curve can be drawn through the points as shown in the figure. Note that the y-axis had to be split up using a range of 1.30 g/L divided by 90 divisions (1.30/90 = 0.014 g/L per division). For convenience this ratio is rounded to 0.02. This means that each division represents 0.02 g/L so every five divisions corresponds to 0.1 g/L. Sometimes, it is necessary to split up the divisions so that they represent some fraction rather than a whole number. In the plot given in Fig. A-4 the point at 250 °C and 0.67 g/L was not plotted. For practice plot this point on the graph. The plot of density versus temperature gives a curved line from which we can conclude that the density is inversely proportional to the temperature. We know this because as the temperature increases the density decreases.

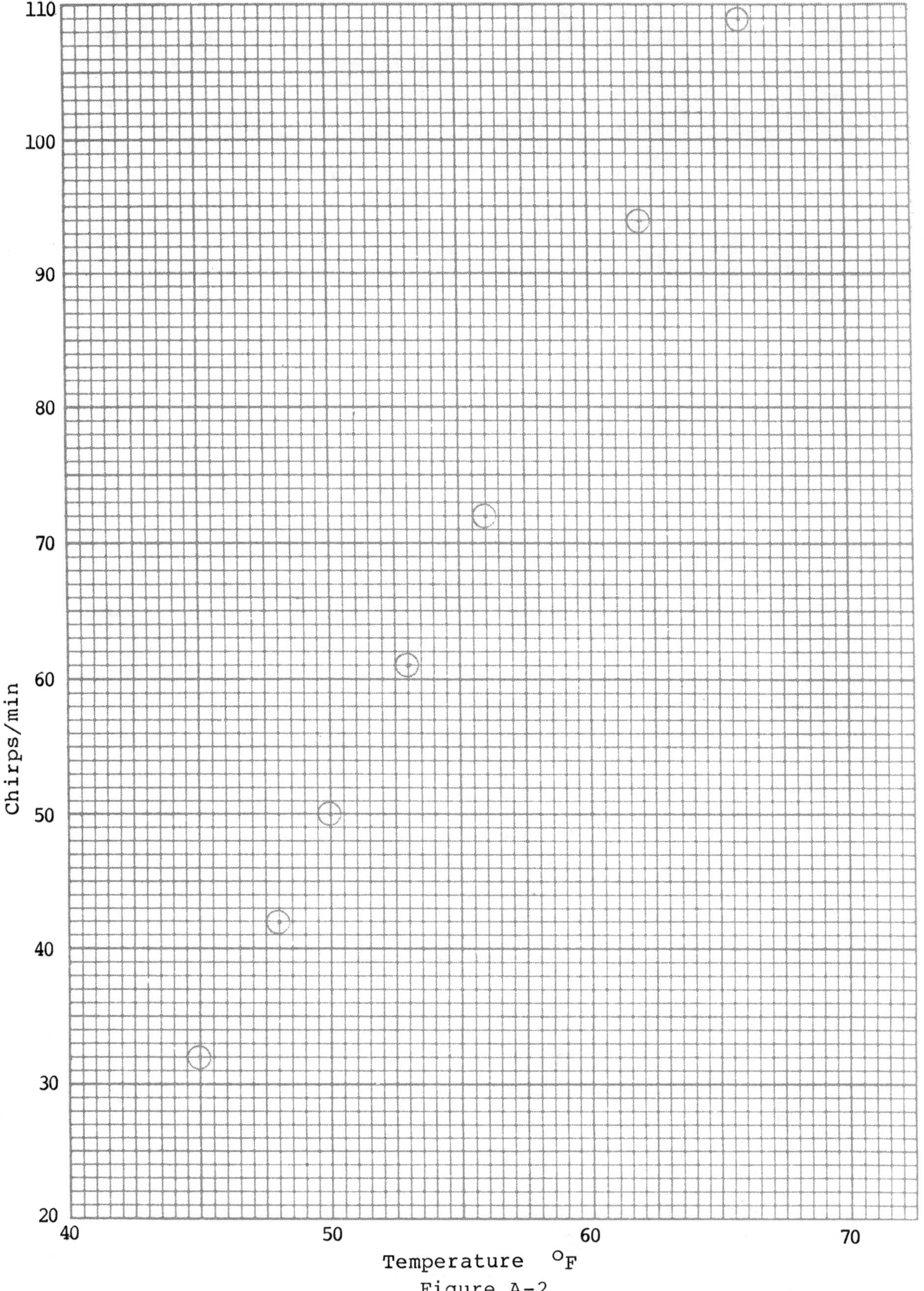

110
100
90
80
70
60
50
40
30
20
Chirps/min
40
50
60
70
Temperature °F

Figure A-2

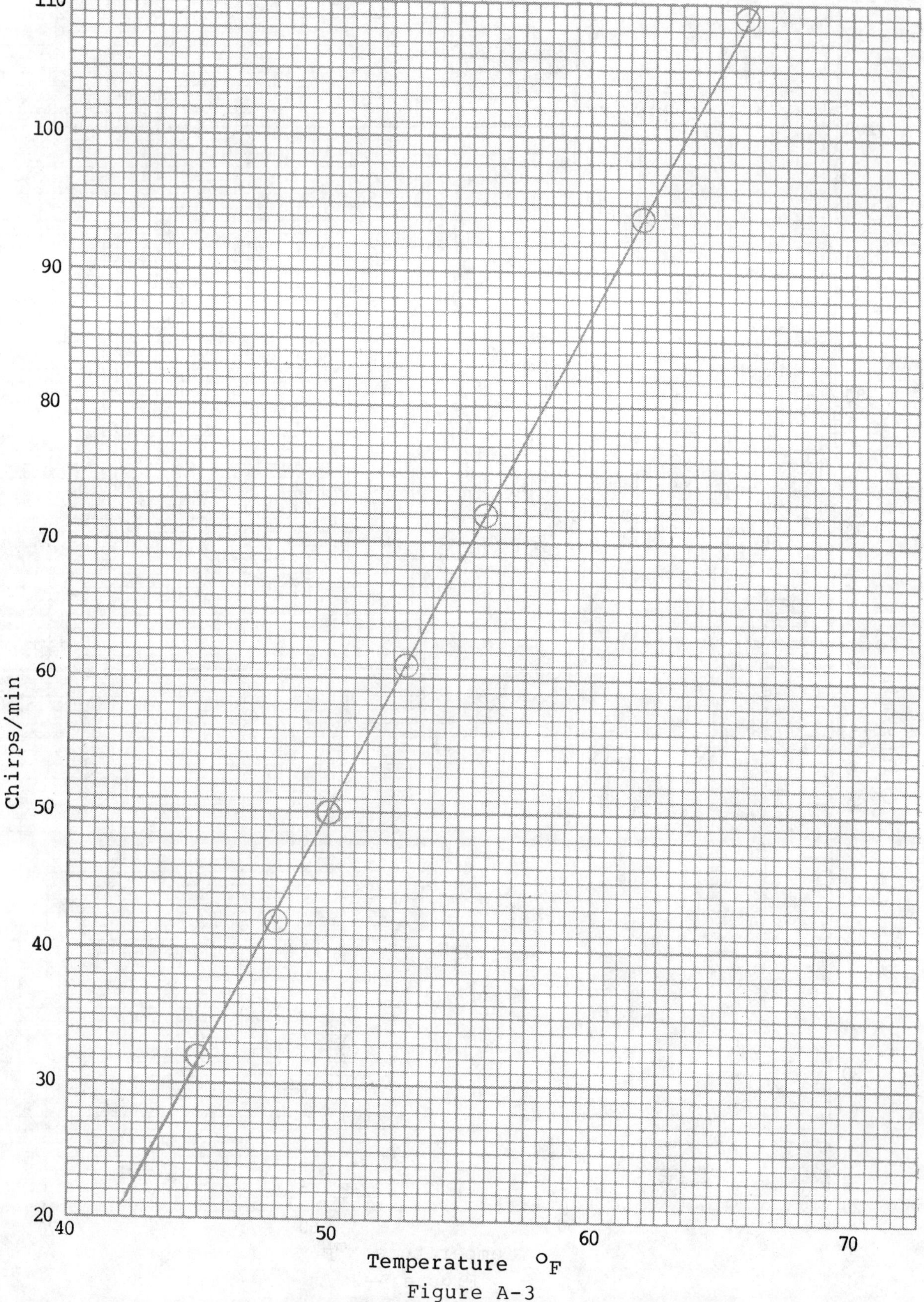
Chirps/min
110
100
90
80
70
60
50
40
30
20
40
50
60
70
Temperature °F

Figure A-3

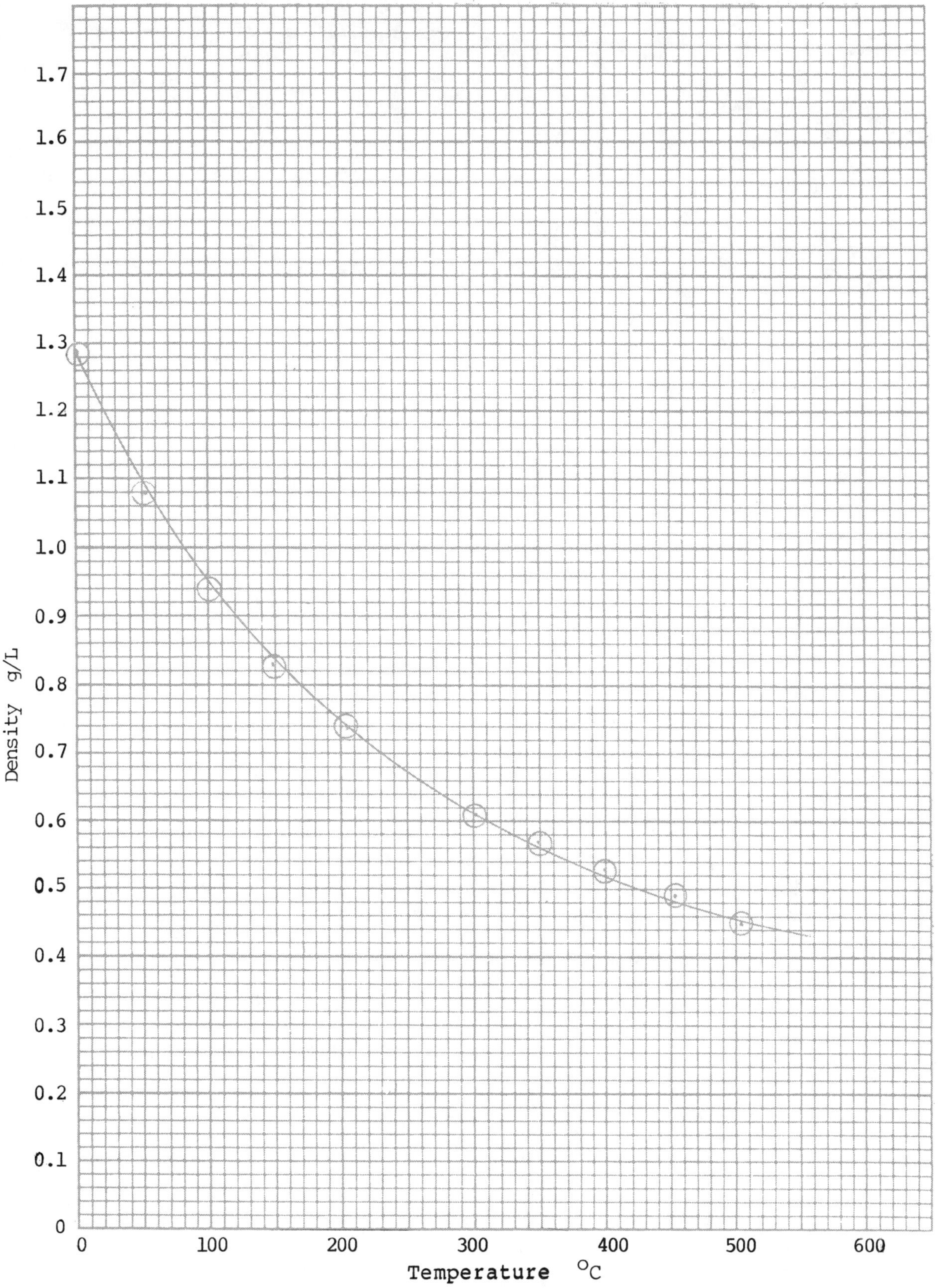
1.7
1.6
1.5
1.4
1.3
1.2
1.1
1.0
0.9
0.8
0.7
0.6
0.5
0.4
0.3
0.2
0.1
0
Density g/L
0
100
200
300
400
500
600
Temperature °C

Figure A-4

Plotting Exercises

A. Plot the data for the vapor pressures of water given in Appendix 1. Plot the vapor pressure versus the temperature.

B. Plot the following data by making a graph of the Fahrenheit temperatures versus the corresponding Celsius temperatures.

°C	0	20	50	65	80	100
°F	32	68	122	149	176	212

Plotting Experimental Data

In this exercise you will collect some experimental data and plot the data on graph paper. For the exercise you will need a laboratory balance, a 10-mL graduated cylinder and a sample of a liquid. You want to measure the masses of various volumes of the liquid and plot the mass in grams versus the volume in milliliters. Collect the data as follows and record your data in the space provided:

1. Weigh a clean, dry 10-mL graduated cylinder to the nearest 0.01 g.

2. Carefully add about 1 mL of liquid, record the volume and weigh the cylinder plus the liquid to the nearest 0.01 g.

3. Carefully add about 1 mL of the liquid to the cylinder along with the liquid that is already in the cylinder. Read the liquid level in the cylinder and record this volume. Weigh the cylinder plus the liquid to the nearest 0.01 g.

4. Repeat step 3 seven more times until you have measured and weighed a total of nine volumes of liquid.

5. To obtain the masses of the samples you will have to subtract the mass of the empty cylinder from each of your weighings.

Mass cylinder ___________________

Mass of sample 1 + cylinder ___________ Volume sample 1 ____________
- Mass cylinder ___________

Mass of sample 1 _____________________

Mass of sample 2 + cylinder ___________ Volume sample 2 ____________
- Mass cylinder ___________

Mass of sample 2 _____________________

Mass of sample 3 + cylinder ___________ Volume sample 3 ____________
- Mass cylinder ___________

Mass of sample 3 _____________________

Mass of sample 4 + cylinder __________ Volume sample 4 ___________
 - Mass cylinder __________

Mass of sample 4 ___________________

Mass of sample 5 + cylinder __________ Volume sample 5 ___________
 - Mass cylinder __________

Mass of sample 5 ___________________

Mass of sample 6 + cylinder __________ Volume sample 6 ___________
 - Mass cylinder __________

Mass of sample 6 ___________________

Mass of sample 7 + cylinder __________ Volume sample 7 ___________
 - Mass cylinder __________

Mass of sample 7 ___________________

Mass of sample 8 + cylinder __________ Volume sample 8 ___________
 - Mass cylinder __________

Mass of sample 8 ___________________

Mass of sample 9 + cylinder __________ Volume sample 9 ___________
 - Mass cylinder __________

Mass of sample 9 ___________________

Summarize your data in the following table and use a piece of graph paper to plot the mass of the liquid samples versus the volume.

	Volume (mL)	Mass (grams)
Sample 1		
Sample 2		
Sample 3		
Sample 4		
Sample 5		
Sample 6		
Sample 7		
Sample 8		
Sample 9